Dysbiosis of the Evolved Intestinal Microbiome

Dysbiosis of the Evolved Intestinal Microbiome

Lessons for Health in Future Generations

David Smith

MDPI

Basel • Beijing • Wuhan • Barcelona • Belgrade • Novi Sad • Cluj • Manchester

Author
David Smith
Independent Microbiome Researcher
Alnwick, UK

Editorial Office
MDPI AG
Grosspeteranlage 5
4052 Basel, Switzerland

For citation purposes, cite as indicated below:

Lastname, A.A.; Lastname, B.B.; Lastname, C.C. *Book Title*. Series Title (optional). Year, Page Range.

ISBN 978-3-7258-1524-1 (Hbk)
ISBN 978-3-7258-1523-4 (PDF)
doi.org/10.3390/books978-3-7258-1523-4

Dedication

To my immediate family, Kathryn Dizon-Smith and Callum Hardy, for their continuing interest and engagement throughout the twists and turns of this investigation, and in memory of my late wife, Violeta Sison Dizon-Smith because, without her, who knows where I might be.

I would also like to dedicate this book to Dr. Michael Moseley (1957–2024). A great communicator, he made us think about the relationship between food and health in the modern world. Sadly, I never had the chance to discuss the microbiome with him.

Contents

List of Figures and Tables

About the Author

David Smith received his PhD from Salford University, UK, in 1980, with a thesis entitled "Ionophoric Benzimidazolones" covering the field of synthetic organic chemistry. He then carried out two years of postdoctoral work at Imperial College, London, UK, working with Professor Charles Rees, FRS, and becoming a Member of the Royal Society of Chemistry and a Chartered Chemist (MRSC, CChem) as well as gaining the qualification of Diploma of Imperial College (DIC). The bulk of his working life from 1984 onwards was spent in the pharmaceutical industry, including the Sanofi group of companies in Alnwick, Northumberland, UK, until retirement in 2013 at age 60.

Following retirement, Dr. Smith continued to seek opportunities to remain valuable to society at large, and it was at this stage that the concept of the intestinal microbiome was beginning to be taken seriously. By 2015, Dr. Smith had developed a novel view on microbiome-related non-communicable disease, and his opportunities for collaboration were markedly increased upon joining NoRCEL, the Network of Researchers on the Chemical Emergence of Life. Alongside members of NoRCEL, across the period 2019 to 2023, Dr. Smith was able to coordinate a series of nine increasingly well-regarded articles, in turn culminating in the production of this book.

Preface

Nothing in Biology Makes Sense Except in the Light of Evolution.—Theodosius Dobzhansky

This book is a summary of work published between 2019 and 2023, updated until the spring of 2024. Having retired with an organic chemistry background, the author brings a novel perspective to this very new, multi-disciplinary field and, accordingly, any mistakes are his alone. The work is designed for any interested persons, both in the general public and among healthcare providers up to, and including, specialists in all aspects of biology and related fields. Accordingly, it is couched in terminology that is as near as possible to general language, even including suggestions on pronunciation for the less familiar words—it is often the case that being comfortable with the pronunciation actually aids understanding. In this context, the term "dysbiosis", derived from "dysfunction" and short for the more logical expression "dysmicrobiosis", represents the failure of the microbiome, and is likely to be the ultimate cause of chronic, non-communicable disease.

Controversially, it must be recognised that there are many other microbiomes associated with different parts of the body and, while some are well described, such as the skin, others, such as the womb or brain, are hotly debated. Nevertheless, it seems that any peripheral microbial communities are secondary to the primary intestinal microbiome, presumably because this has evolved to facilitate what we have referred to as maternal microbial inheritance: the transfer of key microbes from the mother to her offspring during the normal birth process, following apparently accidental "contamination" by faeces. It is important to note that what has been termed the "microbiota–gut–brain axis" applies throughout the vertebrates, so that key microbes may be passed directly on from one generation to the next, either through the medium of the egg, or directly via live birth. In fact, there are a number of ways in which the ideas expressed here differ significantly from the normal interpretation of microbiome content and function. Where these ideas clash with previous assumptions, the reader is urged to keep an open mind and seek consensus among interested parties.

Before proceeding any further, it is necessary to thoroughly analyse the current understanding of microbiome-related biology, illustrating gaps in knowledge and highlighting opportunities for further study, including "opposing" antibiotics and "supplementary" probiotics and prebiotics. In addition, although secondary to the direct study of the microbiome itself, opportunity is taken to reflect on other areas of concern in the modern world, as well as the difficulties of reversing the worldwide trend of steadily increasing disease. This is not a new problem. The first unambiguous case was a precise description of what we now know as hay fever, published in the year 1819, with a total of only 28 cases uncovered in the next nine years across the length and breadth of the British Isles. Earlier examples are harder to detect, but an all-too-human fascination with cosmetics derived from toxic heavy-metal ores may, perhaps, have left its mark in the so-called Venus figurines of the old stone age. If so, it is possible to consider diseases of the immune system, weight gain, and poor mental health as endemic afflictions of that perennially inquisitive creature, *Homo sapiens*.

What is the Intestinal Microbiome?

Cellular components: Of course, the main expression associated with the microbiome is the phrase "good bacteria", the aim being to distinguish them from the pathogenic kind that we have been more familiar with for most of the last century. In reality, the intestine is home to a veritable zoo of relatively simple single-celled organisms, both bacteria and their superficially similar kin, the seemingly non-pathogenic archaea. It also includes more complex entities, the fungi and the so-called protists. Indeed, these latter organisms are the most interesting: a mixed bag of primarily single-celled creatures whose defining characteristic is what they are not—they are neither animal-like, nor plant-like, nor yeast-like (i.e., they are not fungi). Not surprisingly, protists are most often associated with diseases, such as amoebic dysentery, or the mosquito-transferred plasmodium parasite that causes malaria. However, it remains an interesting fact that key components of the immune system include a number of semi-independent single-celled entities, formally termed antigen-presenting cells, but often simply referred to as sentinel cells, that roam around our own body tissues acting as guards.

Opportunity: Consider the possibility that microbial sentinel cells may exist inside fully functioning microbiomes, capable of transferring immune system-based information from a mother to her offspring. Note that similar agents should therefore be found in sustainable wild-type vertebrate animal populations (although may be absent if laboratory-raised), but may be missing from populations suffering from immune system-related non-communicable disease. In anticipation of their eventual discovery, in this book, these so-far entirely hypothetical entities have been termed benegens on account of their theoretical ability to generate benefit, and by analogy with disease-causing pathogens imposing pathology on an infected organism.

Sub-cellular components: Viruses suffer from the same "bad press" as the parasite-like protists, but the viruses indigenous to the microbiome are the so-called phage-type, which exclusively interact with their co-evolved single-celled partners—of which probably the most well-known are the bacteriophage viruses. Significantly, there is a second type of sub-cellular entity whose importance to the microbiome has, perhaps, been inadequately recognised. These are known as plasmids or, more generally, mobile genetic elements, and there can be hundreds of such small, circular DNA units in each bacterial cell. Although there is great interest in the bacterial constituents of the microbiome, even down to the level of species and strain, in all probability it is the collective genome that will be the more important measure of its potential. (Similarly, the collective genes expressed via mobile genetic elements have been termed "the mobilome", but this is currently indistinguishable from the genomic microbiome itself).

Opportunity: Move away from the current investigation of individual bacterial species and strain and move toward assessment of mobile genetic elements with respect to overall microbiome function. Indeed, it could be that the above-mentioned benegens act to coordinate the mobilome, working alongside our own genetic inheritance. Presumably epigenetic (gene-modifying) processes are involved in any microbiome–body interactions, especially during the growth stages.

The role of the intestinal microbiome: Sadly, it seems that the default state of much of modern humanity is what has been termed "dysbiosis", in which a malfunctioning microbiome is associated with a "triple plague" of increasing disease, potentially involving a badly functioning immune system, weight gain, and poor mental health, or in some cases, all three. On the whole, it seems that the pattern of disease reflects what the microbiome ought to do, but that something caused it to break down. Although a great deal of emphasis has been placed on specific substances such as the short-chain fatty acids (SCFAs), perhaps the best way to assess the contribution made to adult health is to compare different populations. While studies are already underway on those peoples who have had very little exposure to industrialised life, often showing both a highly diverse microbiome and low levels of non-communicable disease, there exists a substantial population of what may be termed a control group: adults living within the developed world who have had their large intestine removed by total colectomy. Significantly, it is possible for female colectomy patients to become pregnant and to give birth in either the normal way or by C-section. Whichever method of delivery is chosen, however, if the mother has no intestinal microbes then, of necessity, neither will the child.

Opportunity: There exists an extensive cohort of more-or-less healthy colectomy patients, many of whom may well be enthusiastic participants in a study helping to determine the role of the microbiome in the adult. It is likely that a three-way comparison would provide more knowledge: "dysbiotic" (i.e., normal) populations living within a heavily industrialised environment, "healthy" populations who have had little exposure to such environments, presumably with a functionally intact microbiome, and people whose intestinal microbiome has been physically removed by total colectomy.

Antibiotic resistance: Antibiotics were initially developed when infectious disease seemed to be the greatest threat, while the dangers associated with chronic conditions went unrecognised. However, the emphasis has now shifted significantly to non-infectious conditions, with an increasing realisation of microbiome involvement. Although, in principle, this observation should bring an end to the development and use of antibiotics, in fact, our practical, cultural, and economic dependence on these agents is greater than ever. Interestingly, the way antibiotics work within the body is very different from inside the gut lumen. Once an invader is detected by the above-mentioned sentinel cells, the immune system starts to react, while the pathogen tries to grow faster than our defences can be produced. The antibiotic acts to slow this growth phase, allowing our immune cells to get the upper hand. Once the struggle is over, the lymphatic system proceeds to drain away the debris, allowing the area to heal properly. By contrast, the situation is very different inside the gut lumen, where any post-antibiotic debris merely remains to be incorporated into further microbial growth. It is this point where the problem occurs, as the metabolising power of the microbiome is unleashed on these agents of microbial destruction. Once the genes for antibiotic resistance have been assembled, moreover, they are packaged up as mobile genetic elements and spread around the microbial world, eventually ending up in those more robust entities responsible for infectious disease. In turn, the increase in antibiotic resistance leads to sequential treatment with ever-stronger antibiotics, all too often leading to major problems related to the overgrowth of *Clostridium difficile* (now known as *Clostridioides difficile* or *C. difficile* or, even more simply, as *C. diff*).

Opportunity: Sadly, although there are options to radically change our antibiotic-related way of life—rapid diagnosis, injectable antibiotics, and bacteriophage therapy among them—few can come to fruition without major societal change. In the meantime, however, it is necessary for individuals to follow orthodox medical advice as, at the moment, it is the best that we have got.

Probiotic surprise: The twin ideas of "good" bacteria and "healthy" yoghurt have a long history, so it is not surprising to find that modern consumers had a liking for these products, a fact that eventually led to the extensive industry that we see today. In spite of the advertising, however, what is surprising is that there is little evidence that these hypothetically good bacteria actually end up in the microbiome at all, possibly due to being outcompeted by the resident microbes. Indeed, it seems that many of the bacteria actually occupying the large intestine have not been successfully grown in Petri dishes, perhaps suggesting that their genes have become modified to depend on a symbiotic relationship with the gut wall. Conversely, as probiotics are amenable to such techniques, they may actually be low-key pathogens. Although surprising, the idea that benefit could be provided by immune system exposure to low-pathogenicity microbes is feasible.

Opportunity: Harnessing the placebo effect to cure disease has been a researcher's dream for more than a century and, although they may not thank me for saying so, the probiotics industry may have come close to realising this objective. The expression used in this book is the immune–gut–brain triangle, in which stimulation of the immune system puts both the gut and brain on alert. At best, the right degree of stimulation can tide over (as they say) a difficult period until health is resumed, while overstimulation may give rise to the uncomfortable consequences more closely associated with traveller's diarrhoea. Regardless of the exact description, however, the industry is appreciated by its customers and is likely to remain viable for the foreseeable future.

The modern world: Unfortunately, there are so many high-profile threats to health in the modern world that it is a hard to quantify the overall problem. There is, however, one notable exception, in that it has been repeatedly shown that there is no causative relationship between vaccination and autism. By contrast, microplastics and the so-called "forever" polyfluorinated compounds are genuine unknowns, while the effects of ultraprocessed food and social media, while heavily studied, remain unclear. Although changes in the bacterial microbiome are readily observed under many circumstances, and may be correlated with negative consequences, in themselves these changes are not necessarily causative.

Reversing the damage: As indicated above, this book is predicated on the idea that babies should be born with an intact microbiome picked up from the mother during natural birth. If so, the microbiome can be considered to be a single microbial community working alongside the body in a hand-in-glove relationship, passed on from one generation to the next. In turn, this helps explain its virtual absence from medical/scientific literature prior to the late 20th century. Over time, however, something has changed, with rising levels of disease seemingly following a reduction in microbial diversity. The resultant triple plague epidemic shows no signs of improvement from one generation to the next. Granted that the problem has developed over generations, sadly, it seems unlikely that there will be an easy solution.

As Professor Dobzhansky reminds us, however, a fully functioning microbiome couldn't just have arrived from nowhere. It must have evolved.

David Smith
Author

Acknowledgments

These articles were prepared in collaboration with NoRCEL, the Network of Researchers on the Chemical Emergence of Life: Sohan Jheeta, Miryam Palacios-Pérez, and Kathy McGrath, along with our collaborators, Georgina López-Cortés, Bernadette Street, and Hannya V. Fuentes, on this significant campaign to expand the boundaries of known biology.

There is no better way to say thank you than by laying out our work with full scientific precision:

1. Jheeta, S.; Smith, D. Seeing the wood for the trees: A new way to view the human intestinal microbiome and its connection with non-communicable disease. *Med. Hypotheses* **2019**, *125*, 70–74.
2. Smith, D.; Jheeta, S. Measuring microbiome effectiveness: A role for ingestible sensors. *Gastrointest. Disord.* **2020**, *2*, 3–11.
3. Smith, D.; Jheeta, S. The epidemiology of the dysfunctional microbiome in animals and in humans: The propensity for the development of non-communicable disease. *EC Gastroenterol. Dig. Syst.* **2020**, *7*, 83–93.
4. Smith, D.; Jheeta, S. Microbiome-gut dissociation: Investigating the origins of obesity. *Gastrointest. Disord.* **2021**, *3*, 156–172.
5. Smith, D.; Palacios-Pérez, M; Jheeta, S. The enclosed intestinal microbiome: Semiochemical signals from the Precambrian and their disruption by heavy metal pollution. *Life* **2022**, *12*, 287.
6. Smith, D.; Palacios-Pérez, M; Jheeta, S. Microbiome-gut dissociation in the neonate: Obesity and coeliac disease as examples of microbiome-function deficiency disorder. *Gastrointest. Disord.* **2022**, *4*, 108–128.
7. Smith, D.; Jheeta, S.; Fuentes, H.V.; Palacios-Pérez, M. Feeding our microbiota: Stimulation of the immune/semiochemical system and the potential amelioration of non-communicable diseases. *Life* **2022**, *12*, 1197.
8. Smith, D.; Jheeta, S.; Fuentes, H.V.; Street, B.; Palacios-Pérez, M. Microbiome-gut dissociation in the neonate: Autism-related developmental brain disease and the origin of the placebo effect. *Gastrointest. Disord.* **2022**, *4*, 291–311.
9. Smith, D.; Jheeta, S.; López-Córtes, G.I.; Fuentes, H.V.; Street, B.; Palacios-Pérez, M. On the inheritance of microbiome-deficiency: Paediatric functional gastrointestinal disorders, the immune system and the gut-brain axis. *Gastrointest. Disord.* **2023**, *5*, 209–232.

Thanks are also due to Sarah Hulbert, Laura Wagner and Sherlyn Tshering, and the team at MDPI, for patiently navigating the minefield of font and format; hopes and hyperlinks. In essence: many thanks for their patience and persistence in helping this old "boomer" get his new-fangled ideas into the 21st century!

P.S. It is important to note that, although the open source approach to academic publishing has often been criticised, as long as peer review and due diligence is rigorously applied, it seems that there is no better way to expose genuinely new ideas to public scrutiny.

1. Introduction

While the epidemiology of infectious disease has been raised almost to a branch of mathematics, the epidemiology of non-communicable conditions, taken as a whole, remains an open field.—David Smith

Autism and I: Looking back, it seems that I developed an autism-like condition in high school although, of course, it was not recognised as such in the 1970s. Accordingly, I suspect it will be no surprise to anyone that I became a scientist, gravitating to chemistry and joining the Royal Society of Chemistry of the United Kingdom. Retirement from the pharmaceutical industry allowed me to give informal talks to retiree audiences, focusing on the natural history of the elements, and giving talks to various groups with titles such as "From the Stars to Your Sandwich". Later, I joined the non-funded organisation known as NoRCEL (Network of Researchers on the Chemical Emergence of Life). Founded by Dr. Sohan Jheeta, it encourages open thinking and draws like-minded people into the study of environments potentially conducive to the appearance of life. At this point, all the threads started to come together, and I finally realised exactly how the lack of key microbes within the evolved microbiome could have led to the health conditions that we see around us today, including the one that we now call autism. The ultimate result is this book.

The Development of a Novel Hypothesis: As detailed above, an increasingly well-regarded series of nine articles followed from this initial inspiration, crafted in conjunction with NoRCEL. This book both summarises and extends these works, and espouses what can be best described as a dual inheritance hypothesis: that, for a vertebrate species to survive in the long run, it is necessary to inherit both parental genes and a fully functioning maternal microbiome. This work is based around aspects of both evolution and of the epidemiology of non-communicable disease. Its guiding principle is that the genetic inheritance appropriate to the species is complemented by a second inheritance, the more flexible microbiome, derived from the mother and passing on information about the microbial environment in which she lives. Note that this "calibration" system works by the seemingly accidental faecal contamination of either egg or live birth, as appropriate, and it seems reasonable to suppose that its loss is associated with the manifold autoimmune and inflammation-associated diseases, including obesity, that have been rising in both extent and severity in modern times.

All the Gifts: The "Opening of Pandora's Box" is a theme drawn from Greek Myth. In essence, the eponymous box is said to contain all the gifts, but they turn into curses when the box is opened. Accordingly, the lid is swiftly shut, leaving only one, the gift labelled "Hope", trapped forever. In our world, while technology delivers on all its promises, it also leaves the curse of anthropogenic climate change (global warming), as well as more insidious problems caused by pollution of ground and water supplies. While it is the modern, highly processed, "industrialised" diet that is often held to blame for diseases such as obesity, this work describes how the poisoning of the microbiome leads to the malfunctioning of our body, so it can no longer cope with the sub-optimal food environment in which we find ourselves.

In this analogy, the "Hope" is that technology can fix the problems that it contributed to and, eventually, to repair the microbiome.

Getting Fat: Being initially unaware of the autism-like condition mentioned above, two separate strands came together to initiate this work. The first started around the turn of the millennium when I, purely accidentally, discovered that I was getting fat. Although there is nothing unusual about middle-aged men putting on weight, it occurred to me that it should not really be happening; that it was some kind of symptom-free, undercover disease. Moreover, my organic chemistry background told me very clearly that all the available, strongly promoted explanations of the time must be wrong: that neither saturated fat nor pure, white, crystalline table sugar (sucrose) were inherently deadly. The second strand came much later, following from the fact that, although obesity-related disease was rising all over the world, my employers within the pharmaceutical industry were having a hard time finding profitable drugs to exploit these burgeoning (in every sense of the word) new opportunities. Indeed, at some point it was thought that the "blockbuster" bubble had burst, and that future efforts would, most likely, be aimed at the much less profitable business of merely filling niches. Of course, this was before the discovery of the current crop of anti-obesity GLP-1 agonists, although it is interesting to note that, barring antibiotics, most medicines merely cover the symptoms of disease. In turn, these new agents do not affect the still-unknown reasons for weight gain.

Sneezing: One day, as a teenager in a late-1960s summer, I sneezed. It did not take long to figure out that this was hay fever; that it was perfectly normal, nothing to worry about; and that I was just unlucky. Nearly three decades later, however, in the face of a worryingly steep rise in childhood cases of both hay fever and asthma, David Strachan published an article detailing his observations on the occurrence of autoimmune disease in families. Entitled "Hay Fever, Hygiene and Household Size", it caught the imagination of the public and, perhaps, laid the ground for the later probiotic industry. In the year following Strachan's article, David Barker published an article in the *British Medical Journal* entitled "The Fetal and Infant Origins of Adult Disease", specifically including schizophrenia in his examples of adult-onset disease, but the absence of an accepted mechanism of action meant that his hypothesis, although popular with the public, was not widely accepted by researchers. More recently, it has been noticed that the age of first cancer diagnosis is steadily dropping, so that more people are being diagnosed, with more aggressive cancers, below 50 years of age. Significantly, at about the same time as I noticed my hay fever, I was becoming more involved in my high school work and simultaneously losing my childhood friends, a tendency that I now ascribe to an autism-like condition. Alongside these changes, I also started to notice the intermittent pain of irritable bowel syndrome. Although only a mild form, it seems that I express all the characteristic features of the "triple plagues" of non-communicable conditions: autoimmune disease; a disturbance of mental function; and bowel action disorders that may result in eventual weight gain (although I narrowly avoided actual obesity by careful control of my eating habits). As may be seen in the bulk of the book, these symptoms are associated with a reduction in microbial diversity, both in domesticated animals and in humans, and have been associated with industrialisation. At this point, it is worth considering the exact meaning of the word "microbe".

Microbes and the Microbiome: While bacteria have recently been found in huge numbers in the intestine of apparently healthy people, for most of the last hundred years the word "bacteria" has been synonymous with infection, and with disease. Not unnaturally, therefore, the new bacteria-supplementing probiotic industry employed the kinder term "microbe" in an attempt to avoid any negative connotations. Accordingly, unless otherwise stated, in most publications, the words microbe and microbiome should be taken as referring to bacteria and the corresponding community term, bacteriome, respectively. Unfortunately, mere repetition does not imply evidence, and it could be that the focus on readily detectable bacteria obscures other entities that are, perhaps, more important. Essentially the same arguments apply to microscopic fungi: the so-called mycobiome. Other single-celled entities are also found in the intestine, becoming harder to detect the closer they approach to our own biochemistry. In this work, the word "microbiome" refers to all single-celled entities, their genes, and their associated viruses living within a given area of the body, but excluding outright pathogens, whether readily detectable or not. Interestingly, the absence of all these trillions of intestinal microbes, whatever they may be, does not seem to be a problem, as the loss of both colon and microbiome following radical colectomy does not impart any specific problems. In fact, such people retain fertility and can give birth in the normal way. Accordingly, any "theory of microbiome" should account for the following observations:

1. The epidemiology of childhood-initiated autoimmune disease, such as hay fever.
2. The prevalence of infant-onset bowel disorders, poor mental health, and weight gain.
3. The lack of any specific microbiome-related deficiency disease in those individuals who have undergone total colectomy.

Probiotics: While the hypotheses presented in this book cast doubt on the central role of specified species of intestinal bacteria in the maintenance of health, it is clear that probiotics are indeed beneficial, in some circumstances at least. The suggested reason is that these microbes engage with the immune system and, thereby, strengthen both sides of the so-called gut–brain axis, potentially improving both physical and mental health. It is to be hoped that this insight will provide new opportunities for these agents in the future.

Dysbiosis: The term "smoking gun" refers to strong, if circumstantial, evidence that a crime has been committed by a given individual. In the context of this work, however, the disturbance of the intestinal bacteria known as dysbiosis is the smoking gun, i.e., an indication of a damaged microbiome, rather than the actual cause of the damage itself. The epidemiology of non-communicable disease indicates that dysbiosis was initially caused by heavy-metal pollution, perhaps exacerbated by modern foods, medicines, and medical procedures. Accordingly, there is ample scope for improvement in the future, which will hopefully have a positive impact on the overall health of people around the world.

Benegens: Probably the best way to account for the "early-onset" observations of both Strachan and Barker is to imagine the transfer of microeukaryote agents during the birth process, that they somehow set up both immune system and gut–brain axis for future growth. Indeed, this book aims to document the circumstantial evidence that such agents exist in populations living their traditional lifestyles, as they do not suffer from our characteristic non-communicable diseases. In principle, they could be re-introduced during the birth process, potentially contributing to the health and viability of future populations. Granted that pathogens impart pathology; it may be that the term benegen, "benefit generator", is useful to promote further discussion, a point that will be taken further, especially in Chapter 9. Their difference from the bacterial probiotics lies in their more complex nature, more closely resembling our own body cells, and in their need to be delivered during a brief window of opportunity as part of the birth process itself. Finally, perhaps the best way to understand the role of the fully functioning microbiome is to say that it operates across the generations, and is more involved with the future of the species than of the individual.

Avoiding Exploitation: Of course, the prospect that these benegens may need to be discovered in traditional-living societies that are otherwise unexposed to modern influences raises questions about the potential exploitation of such people. It is important to note, however, that these hypothetical microbial sentinel cells would not be a mineable asset in the same way as rare elements, for example, but that they would need to be tailored to match individual biochemistry before they could be applied to the as-yet-unborn families of people suffering from non-communicable diseases. In principle, mechanisms exist for the compensation of communities providing their traditional knowledge for the benefit of humanity as a whole, and this should be applicable here. It is likely that such peoples would want to be left to live their own life in peace and quiet. Perhaps now will be our last chance to grant them that wish.

Structure of the Book: The overall purpose of this work is to re-interpret the bulk of medical/scientific literature from the perspective of microbiome-function deficiency disease. The form this takes is effectively a super-review, in which recent reviews of seemingly different areas have been assembled into a hopefully relevant whole, although no attempt has been made to be completely comprehensive. After the Introduction, Chapter 2 sets out the hypothesis itself, in some detail but with few references, and Chapter 9 details what needs to be done in the longer-term. The rest of the book is divided into three parts, as follows:

1. Part 1 refers to the 20th century legacy ideas that are still being expressed deep into the present century. The first two chapters differ only in emphasis: while Chapter 3 is largely descriptive, without using the word "microbe" at all, Chapter 4 focuses primarily on gene-based explanations of non-communicable disease, with the words microbiome and microbiota largely being used to indicate unfathomable complexity. As stated above, while microbe-related words are often used in the second chapter, in reality, the quoted authors refer almost exclusively to bacteria and the bacteriome.
2. Part 2 is designed to remind readers that, for all our 21st century technology, we are still creatures with aeons of evolutionary history behind us. Chapter 5 therefore extends ideas of symbiosis into the "newly evolved" vertebrate animals, while Chapter 6 applies these concepts to humanity itself. It is in

this second part that a potential role for benegens to act as a link between the bacteria and our vertebrate body cells is first discussed.

3. Part 3 sets out the consequences for people. Chapter 7 puts microbiome-related disease into the context of the dual inheritance hypothesis, and Chapter 8 focuses on short-term methods of amelioration. Finally, Chapter 9 describes the hoped-for solution, which involves the reintroduction of missing microbes.

2. Between Health and Disease—A New Hypothesis

2.1. *Disease*

While health can be considered to be the absence of ongoing illnesses and physical injury, non-communicable disease may be classed as a disorder of function, of things going wrong. So far, so straightforward—but the advent of genetic testing has complicated the situation in that it flags the possibilities of such diseases that may yet come to pass. We are also plagued with a series of half-ailments, so that, for example, there can be some confusion as to whether a mild dose of hay fever is a full-blown disease, or merely a nuisance. Intellectually, we may know that an allergy is definitely a disease, but it does not necessarily seem that way in the day-to-day sense of the word. Mental illnesses present their own difficulties, as there are rarely any agreed-upon physical measurements for unambiguous diagnoses. Are they really diseases, or merely characteristics? Are people mentally ill, or are they actually healthy but basically evil, a notion that is buried deep in the various cultures of the world? Equally, an otherwise symptom-free gain in weight leaves people perfectly at ease with themselves, and may appear absolutely normal when everybody in their neighbourhood, country, on television or, indeed, across the entire world looks much the same. This "neither here nor there" state is unsettling. Like Schrödinger's cat, neither fully alive nor completely dead, are we healthy or are we not?

2.2. *Health*

Although we tend to forget the fact, there is a general understanding that our physical bodies are the product of long evolutionary processes. Long before the development of our modern understanding of genetics, in the mid-19th century, Charles Darwin identified natural selection as the basis of the origin of species. In turn, these processes depend on the transmission of favoured characteristics across generations and, of course, most research effort has focussed on our genetic inheritance. More recently, however, we have become aware of the importance of the microbiome, the vast number of unicellular entities living all around our body, albeit primarily in the intestine. Since the development of the late-19th century germ theory of disease, there has been a lot of research effort put into the elimination of suspect pathogens but, so far, little thought has gone into the retention of potentially valuable microbes, especially from one generation to the next. In addition to antibiotics, antiseptics and the spread of hospital-based, theoretically sterile caesarean section, further characteristics of modern "industrialised" times include changing types of food and rising levels of pollution. There is ample evidence of the cumulative effect of all these changes on reducing the microbial diversity within our intestine, alongside much speculation of its consequences for the understanding of disease. Whatever the reason, this sharp drop in the variety of microbes living in the intestine is, perhaps, best treated as a deficiency disease: the absence of some essential, but as yet unknown, function of the microbiome. Modern-day scientific understanding is largely based on the concept that an average healthy person will be the same from one generation to the next. Sadly, however, it seems that people are becoming less healthy with time and, indeed, can we really say that anybody is truly healthy? In other words, in the search for what may be characterised as "true health", we should look at individuals from populations that can be shown to be free from non-communicable disease. As described in the following chapters, it seems that such people also retain microbial diversity within their intestine.

2.3. *An Own Goal*

One of the most common healthcare-acquired infections across the world is diarrhoea caused by the anaerobic bacterium *C. difficile* (*Clostridioides difficile*, often known by its earlier name of *Clostridium difficile*). However, modelling suggests that pre-existing, asymptomatic "infection" by *C. difficile* is activated, rather than acquired, within clinical settings, i.e., that patients brought the bacteria into hospital with them [1]. Diarrhoea tends to follow disturbance of the gut microbiome, perhaps implying that *C. difficile* is not so much an infectious agent in itself, and more a particularly tenacious inhabitant of the intestine, a marker of disease as much as its cause [2]. Whether true or not, a key point is that people who have undergone a colectomy, either relating to past *C. difficile* overgrowth or from some other cause, do not seem to have any specific, recognised conditions stemming from the absence of the intestinal microbiota themselves. Indeed, they retain fertility [3] and can deliver their resultant babies with little additional difficulties relative to normal birth [4]. What does this say about the function of the microbiome?

2.4. *A Kaleidoscope of Disease*

As the name implies, non-communicable diseases are those conditions for which an infectious agent has not been observed. However, the process of eliminating potential agents invariably requires a great deal of detective work, and is rarely without controversy. Perhaps of the greatest relevance in the context of this work are the observations of a mid-20th century surgeon, Denis Burkitt, who took his skills around Africa just as it was on the transition between the traditional and modern ways of life [5]. Burkitt identified a number of conditions apparently belonging exclusively to what he called "Diseases of Modern Western Civilization" (Table 1), stating that individuals may suffer from more than one such complaint at any given time. It seems clear that Burkitt emphasised those diseases that he deemed to have a connection to his preferred hypothesis, a lack of dietary fibre, which he linked to obesity, to poor cholesterol metabolism and to high intra-abdominal pressures upon defecation, along with low faecal volumes compared to people living in traditional societies. Surprisingly, he does not mention ischaemic stroke, possibly counting it under the umbrella of cardiovascular disease and, although overt autoimmune diseases are conspicuously present in his "Modern Western Diseases", he clearly states that he has no explanation for their epidemiology. Less surprisingly, however, Burkitt illustrates one serious weakness of 20th century medicine that is still evident today: a tendency to disregard mental health issues.

Table 1. Burkitt's "Diseases Characteristic of Modern Western Civilization".

Appendicitis	Coeliac Disease *	Coronary Heart Disease
Deep vein thrombosis	Diabetes, type 2	Diverticular disease
Gall stones	Haemorrhoids	Hiatus hernia
Multiple Sclerosis *	Obesity	Pernicious anaemia *
Pulmonary embolism	Rheumatoid arthritis *	Thyrotoxicosis *
Tumours of the bowel *	Ulcerative colitis *	Varicose veins

* While many (possibly all) non-communicable diseases have an underlying autoimmune component, those marked with an asterisk are due to an obvious immune system disturbance. Source: Author's compilation based on data from Burkitt [5]; used with permission.

2.5. *A Vicious Circle*

As will be seen, the dual inheritance hypothesis suggests two primary defects: lack of control of the immune system leading to autoimmune disease on the one hand, and breakdown of the gut–brain axis resulting in reduced gut motility on the other. The latter effect works by poor control over the normally well-synchronised gut movements called peristalsis. In turn, this leads to the ill-defined group of conditions commonly called functional gastrointestinal disorders, including the generic term of irritable bowel syndrome. Of Table 1, obesity, stroke, heart disease, type 2 diabetes, and gallstones are rendered more likely as digested food is rapidly over-absorbed into the bloodstream, eventually being deposited as fatty substances, including cholesterol, at dangerously high levels within the body. A consequence of excessive absorption of food is that the microbiome is starved of nutrition, rendering it unable to exert control over gut motility and setting up a vicious circle of continuing disease. Significantly, however, eating foods containing substances that the body cannot readily digest, such as dietary fibre, may ameliorate some of the effects of microbiome-related disease. Two examples help to illuminate the situation.

2.6. *Varicose Veins*

Burkitt divides the diseases of Table 1 into various categories, lumping varicose veins into the "common venous disorders" caused by raised intra-abdominal pressures upon defecation, along with such prestigious companions as deep vein thrombosis, pulmonary embolism, and haemorrhoids [5]. Burkitt quotes the opinion that "over half of all urbanized Western people would develop varicose veins if they lived long enough" and, indeed, granted the likelihood that these conditions can only be produced by raised intra-abdominal pressures, it may be that their presence in an individual is an unambiguous indicator that such diseases are likely to exist within their entire population. In other words, the presence of varicose veins in an individual suggests that the entire population is at risk from microbiome-related disease.

2.7. *Hay Fever and Its Implications*

While Burkitt reported the presence of immune system-related conditions, perhaps it is unlikely that he would have counted the snuffle and sneeze of hay fever as a genuine disease. Interestingly, the first reported case of this condition was described in 1819 by a sufferer, a British medical doctor by the name of John Bostock, who described it as "catarrhus aestivus", or summer catarrh [6]. The main value of Bostock's observation was that he followed it up with nine years of exacting epidemiological investigation. Published in 1828, the results of his studies, across all levels of society and every corner of the United Kingdom, showed that just 28 people yielded unambiguous diagnoses, all from the same well-to-do segment of society that he himself belonged to [7]. Combined with Burkitt's observations and modern-day experience of autoimmune disease, it seems that these non-communicable conditions follow the expansion of industrialisation around the world. It is possible that the initial impetus could be attributed to well-off ladies beautifying themselves with early cosmetics containing toxic metal ions, with deleterious consequences for their children. Although unknowingly, later industrial pollution effectively democratised such disease, especially as the anti-knock agent tetraethyllead was introduced into the petrol burned in cars throughout most of the 20th century [8]. These, and other conveniences of modern life, have probably contributed to the present degraded microbiome and, therefore, to the expanded range of diseases that we are now becoming sadly familiar with. Alongside varicose veins, the presence of autoimmune conditions may well be an indicator of a population suffering from microbiome-related disease. Furthermore, the logic of a degraded gut-brain axis implies a causative role for mental health conditions, and it could be that, in the absence of infection or poisoning, the appearance of any individuals exhibiting mental health issues also implies microbiome-function deficiency throughout that population. The current status of psychiatry has been summarised in a recent book by the sociologist Andrew Scull "*Desperate Remedies: Psychiatry and the Mysteries of Mental Illness*", illustrating the intermittent, placebo-like success of either "talking" or "chemical" cures [9]. Although Scull covers the early germ theories of mental illness, neither nutrition, microbiome, nor the gut-brain axis are emphasised. Accordingly, it may be that his work usefully defines the end of one era, and the beginning of another. Only time will tell.

2.8. The Nature of the Microbiome

On the whole, the fully functioning microbiome is best considered as a single unit, a community of microbes cooperating with the host via gene-led functions, rather than merely as a collection of individual bacteria. In essence, it is an evolved entity, a co-factor working alongside multicellular animals. Accordingly, in this book, there are no lists of specific "good" bacteria, such as may be associated with probiotics, for example. Instead, microbial diversity is key, whether it refers to bacteria, whose value to the microbiome is to produce chemicals, or the more complex benegens perhaps more closely associated with microbiome functionality. These latter cells are called eukaryotes (pronounced "you-carry-oats") from their resemblance to a nut containing a kernel, the "eu-" prefix implying "good" to distinguish them from simpler prokaryotes such as the bacteria.

2.9. Sentinel Cells

In principle, the DNA-based lifeforms that exist around us today can be traced back to some point shortly after the beginning of the Earth itself. This is not to say that there were never any other forms of life, just that the current gene-led system proved to be the most successful. Initially, it seems, there were two "domains" (i.e., types) of single-celled lifeforms, the prokaryote Bacteria and a similar group of organisms that have been named the Archaea (domain names are capitalised). Representatives of these two domains later merged, giving rise to the Eukarya, the third domain of single-celled lifeforms, which later evolved into multicellular entities. In the recent past, of course, all bacteria were classed as pathogens, while independently living unicellular eukaryotes, even those not associated with disease, may still be referred to as parasites. However, there are similar entities endlessly patrolling our body tissues as part of the immune system and, as these are the first line of defence against interlopers, many of them fall under the general term of sentinel cells, such as dendritic cells [10]. The key point is that these sentinel cells have the ability to physically move an antigen-containing component from the site of its discovery to the site of antibody production.

2.10. Microbial Sentinel Cells

One of the major themes of this work is that immune system sentinel cells ought to have microbial partners patrolling the intestinal lumen, interacting with our bodily sentinel cells in the gut wall so as to tag antigens as friend or foe. If so, a role of these eukaryotic microbial sentinel cells would be to transfer antigenic information from the mother to her offspring during the process of natural birth, thereby effectively calibrating the immune system of the neonate against the up-to-date microbial environment of the mother. Indeed, it is these hypothetical entities that are referred to as benegens—agents that provide benefit—especially in Chapter 9. While the rationale for such entities is clear, their detection will be not be easy as, unlike bacteria or fungi, their biochemistry is likely to make them very difficult to distinguish from the "background noise" of sloughed-off dead cells present in the intestine. In addition, it is possible that these entities simply do not exist in populations suffering significant levels of non-communicable disease. In turn, this necessitates studies in those peoples living their traditional lifestyles, ideally in those populations known to be free from such diseases. As mentioned in Chapter 9, in addition to direct detection, observing the transfer of some sort of labelled antigen from mother to neonate would provide proof of principle, albeit probably in animals at first.

2.11. The Poisoning of Microbial Sentinel Cells

A second major theme of this book is that microbiome poisoning occurs as these cells engage with the microbe-sized heavy-metal-containing particles associated with industrialisation. Indeed, in this hypothesis it is the disappearance of these microbial sentinel cells that gives rise to the malfunctioning immune system invariably seen in cases of non-communicable disease. The third theme is that such microbiome poisoning is not necessarily seen in the affected individuals themselves, but rather in their children and, accordingly, each new generation is more severely affected than the last. Sadly, so far there is no evidence that the microbiome can repair itself.

2.12. Microbiome-Related Terms and Concepts

There are various molecular entities which, while they are not normally classed as "life", act on both pro- and eukaryotes: viruses are small, numerous, and tend to impose themselves upon specific target cells; while mobile genetic elements are relatively small, often circular, portions of DNA, carrying the coding required to make the necessary proteins to carry out various actions. Useful terms are expanded upon below:

1. Microbiome: A complex mixture of entities, not just bacteria, occupying areas of the body while working together (i.e., in mutualistic symbiosis) with their host. Note that the "-ome" expression can be used for different purposes. Thus, for example, the exclusive study of bacteria within an area of the body relates to its bacteriome. The same expression may be used for viruses, "virome", or for fungi, "mycobiome". Logic suggests that the more complex microeukaryotes somehow act as a link between the prokaryotes and our body cells, so that the overall intestinal microbiome acts as a bridge between body and environment (of course, such unicellular eukaryotes could constitute a "microeukaryome"). In this respect, one role of the prokaryotes is to express mobile genetic elements that may be brought in with food. This is an example of another term, the "mobilome", i.e., genes that can be shuffled around, both within and between cells, in what has been called horizontal gene transfer. More generally, the expression "microbiome" refers to the collective genomes and communities of microorganisms and their associated mobile genetic elements, including bacteria, archaea, fungi (yeasts), protists, and viruses, that reside in a given environment.
2. Holobiont and hologenome: The term holobiont—the whole biology—indicates that assemblages of individual species may co-evolve if they collectively confer advantage. In a similar fashion to the term microbiome, the related terms holobiont and hologenome could be considered to be essentially equivalent. Although the term holobiont was employed by Lynn Margulis in her studies of symbiosis [11], it bears some similarity to the unrelated concept of holistic medicine, that the body should be treated as a whole. While a laudable concept on its own terms, of course there is nothing that cannot be put to bad use, and any of the so-called "alternative" medicines can be used as vehicles for conspiracy theories [12]. Importantly, however, it is worth noting that "holistic" is the opposite to reductionist, analytical, approaches, where organs are looked at in isolation. As we now have the ability to grow mini brains [13], for example, the missing input from the gut-brain axis should be considered and, accordingly, observed results may not represent the whole system.
3. Mobile genetic elements: Of course, genes are normally considered to be fixed by gene transfer from parent to child at conception, passing on characteristics such as height or eye colour, for example. However, among the single-celled entities (pro- or eukaryotes) genes may be transferred directly from external sources as part of the overall mobilome of the individual or environment. These

two characteristics are referred to as "vertical" and "horizontal" gene transfer, respectively. Significantly, this effect confirms that any two individuals, even if they differ in terms of their microbiome composition, may nevertheless express the same genetic functionality. As an example, the genetic ability to resist antibiotic action may be spread across the wider microbial environment.

4. Viral shunt mechanism: Initially evoked to explain the retention of organic matter in specific layers within the sea, the principle is that viral (so-called phage) lysis of microbial cells affords dissolved nutrients which, in turn, stimulates the growth of different microbes. In this way, an assemblage of microbes in an enclosed environment, such as the intestine, represents a dynamic situation capable of rapid response to ecosystem change. Accordingly, the intestinal microbiome of a vertebrate animal possesses a flexibility that is complementary to the stability of the evolved multicellular organism itself.
5. Semiochemical: In the broadest sense, semiochemicals are signalling molecules. While hormones relate to messages spread within the body through the bloodstream, pheromones refer to communication between different individuals of the same species. In this book, however, the word semiochemical is used for mutually intelligible messages passed between individual organisms of widely different species, in this case the vertebrate body and microbial community. While the actual chemical may be the same molecule as used by the body—for example, dopamine—its significance will depend on the location and timing of their production. Instead of acting directly on the brain, semiochemicals may simply activate the gut wall, messages being passed on to the brain via the vagal nerves of the so-called gut–brain axis. As illustrated in Figure 1, in the dual inheritance hypothesis it is the balance between microbe-generated semiochemicals and bodily hormones that drives peristalsis, allowing the partition of nutrition so that both components of the holobiont survive. In principle, this system is capable of precise control, so that famine, with its threat to both components of the holobiont, will increase hormone levels and slow peristalsis until the danger passes. While weight can certainly be gained, obesity, with its corollary of non-communicable disease, is avoided.
6. A dual inheritance: A consequence of the holobiont concept is that significant microbial guests must be transferred from one generation to the next. Of course, natural birth is a messy, traumatic process and, indeed, the apparently accidental faecal contamination of the baby is probably the most successful arrangement available. When successful, alongside parental genes, this process allows transfer of the mother's (maternal) microbiome directly to its offspring, be it via egg or live birth.
7. Handshaking and the immune–gut–brain triangle: Interestingly, there is a high level of overlap between the concepts of biology and of computing (see the "triple-layer model" below). One expression used in the latter field is the phrase handshaking, which refers to the exchange of the protocols of communication prior to full merger of two different devices. It is likely that something similar happens between the gut wall of the neonate and the incoming microbiome just after the birth process. Once there, these microbiota drive this handshaking process to both establish the immune system, probably via maternal sentinel cells, and initiate growth of the gut–brain axis, ready for later semiochemical

production. Accordingly, the gut–brain connection has recently been referred to as the microbiota–gut–brain axis, but such is the importance of the three-way interaction with the immune system that the expression "immune–gut–brain triangle" is probably more helpful, as illustrated in Figure 2.

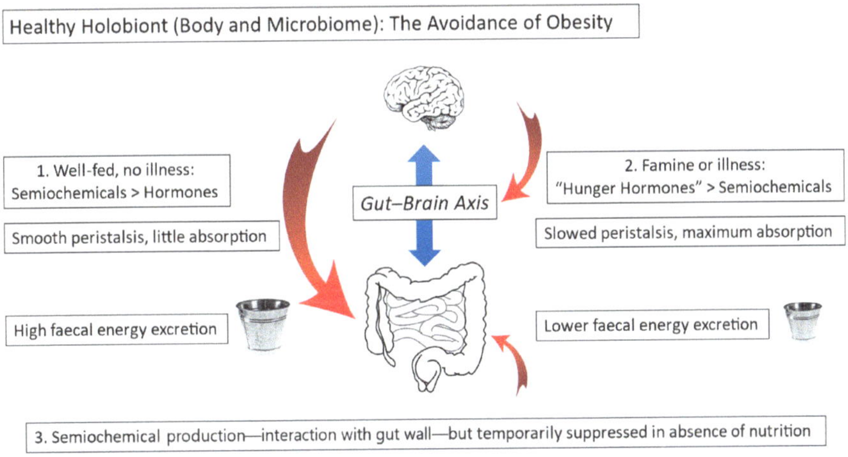

Figure 1. The avoidance of obesity. *Left-hand side*: Indicates health in the body and microbiome, already well-nourished and with adequate food intake. Box 1. In these circumstances, semiochemicals are unconstrained by bodily hormones and activate the gut wall. In turn, these stimulate smooth peristalsis (left-hand red arrow) and allow growth of the microbiome. In due course, defecation removes excess energy by excretion of bacteria-rich faeces. Box 2. When the otherwise healthy holobiont is faced with famine and/or illness, chemical messages, described here only as "hunger hormones", deactivate the gut wall, rendering it insensitive to semiochemicals and slowing peristalsis until the danger passes (right-hand red arrow). In turn, during this period, reduced microbial growth leads to lower levels of faecal energy excretion. Box 3. In this scheme, semiochemical production acts only on the gut wall (lower red arrow), without any direct action on the rest of the body. Note that semiochemical production can be temporarily suppressed in the absence of nutrition. By contrast, a malfunctioning microbiome at birth may not activate the gut wall to receive semiochemical signals, leaving it permanently unable to coordinate peristalsis. In this way, we see the pain of irritable bowel syndrome and/or the gain of weight characteristic of modern populations. Source: Adapted from Smith et al. [14], used with permission.

8. Triple plagues: While the term "parental genetic inheritance" is widely understood, its microbial equivalent is not normally considered. This book is based on the view that the transfer of key microbes happens more or less automatically following natural birth, only being noticed when something goes wrong. In this way, we get the triple plagues of immune system, mental health, and weight gain-related gut symptoms akin to Burkitt's "Diseases of Modern Western Civilization". Sadly, once functionality is lost, it is not automatically recovered and, accordingly, it seems that the child's descendants also suffer. In

other words, non-communicable disease tends to multiply down the generations in a fashion that has been termed a snowball effect.

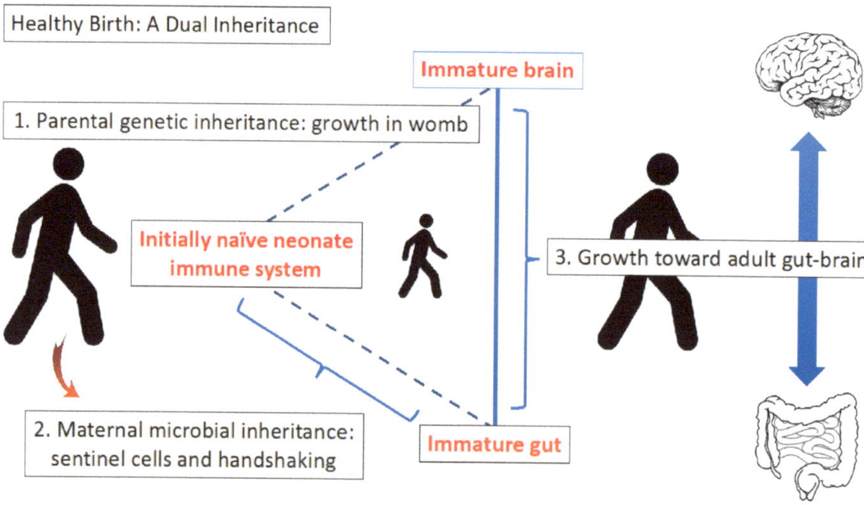

Figure 2. A dual inheritance. Box 1. The foetus grows in the womb according to its parental genetic inheritance, producing an immature brain and gut, along with initially naïve immune system (red writing). Box 2. Birth is accompanied by contamination by the maternal microbiome, in which microbial sentinel cells calibrate the immune system of the infant against the microbial environment experienced by the mother, and other constituents initiate the handshaking process by which the gut wall interacts with the semiochemicals of its intestinal microbiome. Box 3. All being well, the brain and gut grow together, connected by the gut–brain axis. By contrast, a failure of maternal microbial inheritance leads to intermittent gut–brain connectivity and the gut and mental health problems characteristic of modern populations. Source: Adapted from Smith et al. [14], used with permission.

9. Infection versus beneficial inheritance: Of course, we are more familiar with the concept of microbial infection rather than microbial inheritance. While it requires a high pathogen load to overcome immune system defences, it could be that beneficial microbes can be taken into the neonate with a much lower microbial loading (i.e., a high "beneficial virulence"), especially when combined with breastfeeding. In other words, uptake of co-evolved beneficial agents may actually be hard to completely prevent, even when the baby is delivered by C-section. Interestingly, of course, the term caesarean section pre-dates the concept of a sterile operating theatre. Indeed, it could be that non-hospital emergency deliveries actually allow the transfer of beneficial agents more readily than under planned conditions.
10. Dysbiosis and probiotics: The term dysbiosis refers to the concept that many of the ills of modern life may be due to deficiencies in the bacterial composition of the microbiome. This is an early assumption that has neither been confirmed nor denied, and yet it is the basis of the flourishing probiotic industry, essentially treating microbes as inanimate pharmaceutical agents. In this book, the term

dysbiosis is used to indicate a lack of microbiome gene-based functionality, essentially regardless of the precise composition of the microbiome itself, whether of bacteria or of other single-celled entities.

2.13. *A Cellphone Analogy*

As mentioned above, bearing in mind the overlap between aspects of computing and biology, a comparison of cellphones with the functioning microbiome may be useful. Mobile phones have "evolved" significantly throughout the early 21st century, both in make and model, but they all operate within the same overall environment. Just as microbes exchange genetic abilities "horizontally", i.e., directly from one individual single-celled entity to another, so messages can be passed directly from one cellphone to another, more or less regardless of make or model. Interestingly, essentially all human organisations, be they companies or states, attempt to exert control over information flow by use of a firewall. It could be that the microbial analogy of firewall-like control is the function of unicellular eukaryotes.

A Triple-Layer Model: The mobile phone analogy discussed above implies "layers" of function, perhaps as follows:

1. Mobilome: The applications (apps) used by cellphones could be considered to be analogous to the mobile genetic elements (MGEs) transferred between cells. Both apps and MGEs work within the operating system of the cellphone or cell, respectively, and neither can readily be accessed externally.
2. Prokaryotes: Both cellphones and bacteria are plentiful in their respective environments, and both have the capability of uploading operating instructions from an external source. Here, the analogy breaks down a little because, while mobile phones tend toward monopoly-like uniformity, microbiome effectiveness seems to rely on high bacterial diversity, presumably because of the relative difficulty of handling the requisite DNA-based machinery.
3. Microeukaryotes: Organisations usually require a high degree of control over their cellphone networks and the same will apply to the microbial community that is the microbiome. As recent times have seen a dramatic loss of microbial diversity, it is likely that non-communicable disease arises from the loss of microbiome effectiveness, as vital functions can no longer be deployed. Nevertheless, it is most likely that firewall-like control ought to be exercised by the hypothetical benegens that may behave as microbial sentinel cells. In order to be sure, however, we may have to examine populations that do not suffer from non-communicable disease, i.e., the sorts of populations that Denis Burkitt had access to. As will be seen in the text, although threatened, such peoples do still exist.

2.14. *Post-War Generations*

As mentioned above, there is a progressive loss of key microbes as one generation follows the other. The middle of the 20th century witnessed the double European/Asian tragedy of the Second World War. Needless to say, this traumatic event disrupted the normal flow of generations, leading to an increase in post-war births that has been called the baby boom. The idea of naming subsequent generations has caught on, and was followed by a more-or-less consistent succession of generations: Gen X, Gen Y (millennials), and Gen Z, who were themselves having babies as this book was being written. Naturally, there has been an assumption that all such generations are alike and, therefore, that the newer generations can be judged by the standards of the previous one. Sadly, the snowball effect of increased disease referred to above makes this a dangerous assumption, and the younger generations are progressively more prone to previously unremarked-upon illnesses. The connected nature of these conditions is illustrated in Figure 3.

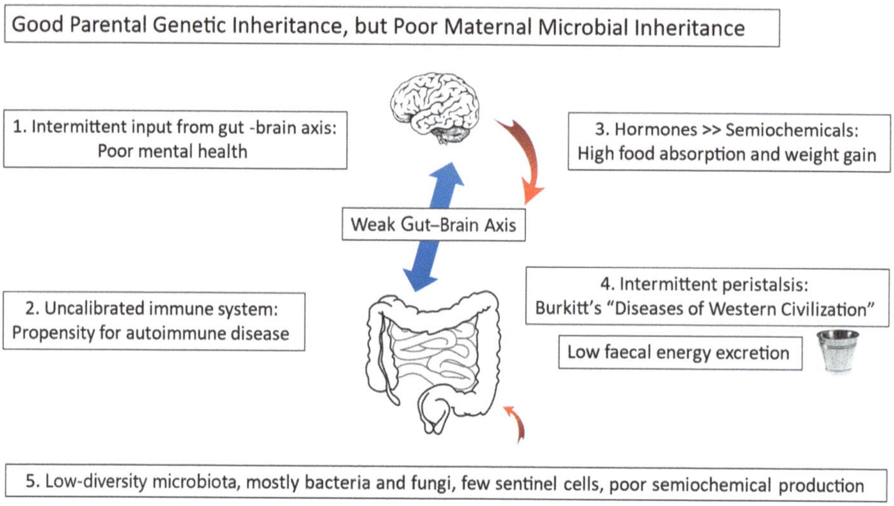

Figure 3. Poor maternal microbial inheritance. Box 1. It seems reasonable to suppose that the brain grows according to the input received. Accordingly, intermittent input from a weakened gut–brain axis leads to distorted brain growth and, therefore, an increased propensity for poor mental health. Box 2. As the immune system remains naïve, it overreacts to harmless stimuli such as pollen grains or food proteins, leading to potentially catastrophic autoimmune disease. Box 3. Poor peristalsis leads to weight gain due to overabsorption of highly processed, readily digested foods. Box 4. In combination with autoimmune disease and weight gain, intermittent peristalsis leads to the conditions described in Table 1, along with the low faecal volume described by Burkitt [5]. Box 5. Low microbial diversity, dysbiosis, is a characteristic of modern Westernised societies in which the bulk of research focuses mainly on bacteria and fungi. As yet, microbial sentinel cells remain undiscovered, and may not exist in societies suffering from non-communicable disease. Likewise, it is likely that semiochemical production remains low in such societies, at least in comparison with people living their traditional lifestyles. Source: Adapted from Smith et al. [14], used with permission.

2.15. Ingestible Sensors

In principle, the functional effectiveness of an intestinal microbiome can be examined by means of an ingestible sensor—a device that, after swallowing, allows measurement of the intestinal concentration of key molecules produced in response to food or other stimuli. Likely following initial trials in larger animals, the ultimate aim would be to indicate the essential microbial function for an effective microbiome, either to confirm its adequacy in the mother, or to be created as a supplement to be delivered to a baby at the time of birth. However, it may be that even a supplemented microbiome is ineffective in the absence of critical microeukaryotes. Only experiments will tell (see Chapter 9 for more details).

2.16. Summary

The hypothesis described in this work is outlined below:

1. Holobiont: The intestinal microbiome of the vertebrate animals is, in essence, a co-evolved microbial community living within its multicellular host. On this hypothesis, it is likely to operate on three levels: microeukaryotes act as coordinators of prokaryotes and as intergenerational sentinel cells; the prokaryotes constitute the majority life-form, constantly changing microbiome composition via viral shunt under phage action; and, finally, mobile genetic elements can be uploaded in response to changes in the outside world.
2. Inheritance: Both parental genes and the maternal microbiome must be passed on to the next generation in a dual inheritance: the foetus grows in the womb or egg according to genetic instructions; and the intestinal microbiome is passed on by apparent contamination as the egg descends through the cloaca, or the neonate through the birth canal, itself adjacent to the anus. Subsequent action is twofold: antigen-presenting, intergenerational, microbial sentinel cells calibrate the immune system of the chick/neonate, and growth of the gut–brain axis is stimulated.
3. Adulthood: At this stage, the fully functioning gut–brain axis can adjust the rate of peristalsis, and in this way, the brain and body can share nutrition with the microbiome. In turn, this action ensures that both components survive through to successful pregnancy more-or-less regardless of potentially unfavourable circumstances. At this point it is worth noting that all of the traits, behaviours, and characteristics of an individual can be classed as its phenotype, which is normally considered to be the effect of the environment upon its genetic inheritance. While this is straightforward in principle, the details for individuals can be complex, as in Crusio's analysis of the curious "phenotype-free mouse" phenomenon, for instance [15]. On the whole, it is probably better to leave this area to the specialists, and simply consider the maternal microbiome to be part of the overall environment in which an individual develops.

References

1. Lanzas, C.; Jara, M.; Tucker, R.; Curtiss, S.; CDC Modeling Infectious Diseases in Healthcare Program (MinD). A review of epidemiological models of *Clostridioides difficile* transmission and control (2009–2021). *Anaerobe* **2022**, *74*, 102541. [CrossRef] [PubMed]
2. Sehgal, K.; Khanna, S. Gut microbiome and *Clostridioides difficile* infection: A closer look at the microscopic interface. *Ther. Adv. Gastroenterol.* **2021**, *14*, 17562848211994736. [CrossRef] [PubMed]

3. Druvefors, E.; Myrelid, P.; Andersson, R.E.; Landerholm, K. Female and male fertility after colectomy and reconstructive surgery in inflammatory bowel disease: A national cohort study from Sweden. *J. Crohn's Colitis* **2023**, *17*, 1631–1638. [CrossRef] [PubMed]
4. Baggio, S.; Pomini, P.; Zecchin, A.; Garzon, S.; Bonin, C.; Santi, L.; Festi, A.; Franchi, M.P. Delivery and pregnancy outcome in women with bowel resection for deep endometriosis: A retrospective cohort study. *Gynecol. Surg.* **2015**, *12*, 279–285. [CrossRef]
5. Burkitt, D.P. Some diseases characteristic of modern western civilization. *Br. Med. J.* **1973**, *1*, 274–278. [CrossRef] [PubMed]
6. Bostock, J. Case of a periodical affection of the eyes and chest. *Med. Chir. Trans.* **1819**, *10*, 161–165. [CrossRef] [PubMed]
7. Bostock, J. Of the catarrhus aestivus or summer catarrh. *Med. Chir. Trans.* **1828**, *14*, 437–446. [CrossRef] [PubMed]
8. Needleman, H. The removal of lead from gasoline: Historical and personal reflections. *Environ. Res.* **2000**, *84*, 20–35. [CrossRef] [PubMed]
9. Scull, A. *Desperate Remedies: Psychiatry and the Mysteries of Mental Illness*; Penguin Books Ltd.: London, UK, 2022; ISBN 9870241509421.
10. Banchereau, J.; Briere, F.; Caux, C.; Davoust, J.; Lebecque, S.; Liu, Y.-J.; Pulendran, B.; Palucka, K. Immunobiology of dendritic cells. *Annu. Rev. Immunol.* **2000**, *18*, 767–811. [CrossRef] [PubMed]
11. Margulis, L. Symbiogenesis and symbionticism. In *Symbiosis as a Source of Evolutionary Innovation: Speciation and Morphogenesis*; Margulis, L., Fester, R., Eds.; MIT Press: Cambridge, MA, USA, 1991; pp. 49–92.
12. Fournier, V.; Varet, F. Conspiracy beliefs and intention to use conventional, complementary and alternative medicines: Two vignette studies. *Br. J. Health Psychol.* **2023**, *29*, 333–350. [CrossRef] [PubMed]
13. Eichmüller, O.L.; Knoblich, J.A. Human cerebral organoids—a new tool for clinical neurology research. *Nat. Rev. Neurol.* **2022**, *18*, 661–680. [CrossRef] [PubMed]
14. Smith, D.; Jheeta, S.; López-Córtes, G.I.; Fuentes, H.V.; Street, B.; Palacios-Pérez, M. On the inheritance of microbiome-deficiency: Paediatric functional gastrointestinal disorders, the immune system and the gut-brain axis. *Gastrointest. Disord.* **2023**, *5*, 209–232. [CrossRef]
15. Crusio, W.E. My mouse has no phenotype. *Genes Brain Behav.* **2002**, *1*, 71. [CrossRef] [PubMed]

Part 1. Gene-Focused Approaches to Disease

Millennial Changes: The first part of this book consists of two chapters covering the state of the medical/scientific world from the perspective of the early 21st century. Westernised medicine developed from medieval metaphysics and religion in two directions, roughly translatable as physics and psychology. Accordingly, for example, high blood pressure could be seen as physics, the body going wrong in a mechanical fashion, whereas the increasingly prevalent "white coat" hypertension, although nearly as damaging as standard hypertension, clearly has a psychological component. This new era of non-communicable disease will see a shift to the dominance of developmental brain–body conditions, as in, for example, the recently highlighted relationship between autism and joint hypermobility and, indeed, one could also predict that other "idiopathic" joint-related conditions, such as rheumatism, could be linked to neurodevelopmental disease. More generally, the following two chapters attempt to document the transition between primarily behaviour-led explanations of the origin of such disease to an emphasis on genes and, eventually, the much-misunderstood microbiome. In this context, it is interesting to note that a common expression is "genes and environment", but without specifying the way in which the two interact. Indeed, it seems that the best way to visualise the role of the microbiome is as part of this process, of the interaction of our parental genetic inheritance with the environment into which the child will be born. Accordingly, subsequent chapters focus on the evolutionary background and on the dual inheritance hypothesis itself, with an aim of both amelioration and, eventually, the prevention of future disease.

3. Non-Communicable Conditions: Background and Attitudes

3.1. Behaviour

Owing to the human drive to understand and explain, it is rare that phenomena are simply described without any attempt at rationalisation. Accordingly, this chapter primarily refers to pre-microbiome explanations of phenomena, with their primary emphasis on behavioural explanations of the more obvious conditions such as weight gain. This theme is illustrated by the 200th anniversary of the inauguration of the *New England Journal of Medicine*, which was celebrated with a 2012 review covering the changing causes of death since their first, 1812, edition [1]. With the exception of the change of emphasis between cannonballs and thermonuclear weapons, there were many similarities over the years, at least until the establishment of the germ theory of disease. Accordingly, the 1912 issue represented a turning point, finally showing a sanitation-led decrease in deaths due to infectious disease. Sadly, however, subsequent editions emphasised the arrival of non-communicable conditions: a steady rise in deaths due to cancer, stroke, atherosclerotic heart disease, type 2 diabetes, neuropsychiatric conditions, and suicide, with the latest, 2012, edition predicting the (eventually observed) drop in average life expectancy. Interestingly, the 1912 edition also prophesied an upcoming epidemic of obesity, blamed on the plethora of new-fangled energy-saving devices such as lifts and, in particular, the ubiquitous automobile [1]. Of course, the characterisation of weight gain as a consequence of labour-saving devices stresses the fact that, initially at least, it is the rich that tend to suffer from such non-communicable conditions, a theme further emphasised below.

3.2. Non-Communicable Disease

Technically, the term non-communicable disease represents health-related conditions that are not caused by an infectious agent. Needless to say, it can be hard to tell if this agent simply has not yet been found and, in a sense, there is a permanent search for the critical pathogens. In the definitive absence of such a cause, however, and in the absence of faulty genes, the main alternative is a functional deficiency, due to either an inadequate diet or poisoning knocking out some vital enzyme, for example. Normally, these conditions are defined by the trouble they cause, and they are often attributed to behavioural problems. For example, a recent review of risk assessment questionnaires (i.e., covering the perceived risk of an individual developing chronic, long-term, non-communicable disease) picked four such conditions as the cause of 71% of worldwide deaths: cardiovascular diseases; cancers; diabetes; and chronic respiratory conditions. The authors reiterated the general belief that most of these are bad habit, behaviour-related, theoretically preventable so-called "lifestyle" diseases, the primary causes being self-administered toxins such as tobacco and alcohol, along with poor diets and a lack of physical exercise [2].

3.3. *Lifestyle*

Of course, the difficulty common to all scientific–medical fields is the obtention of reliable data. Accordingly, the field is wide open to what Casazza, and colleagues, call either "presumptions" (views held in the absence of actual evidence) or "myths" (views so strongly held that they persist in spite of evidence to the contrary). While their 2015 article examines the evidence surrounding no less than nineteen of these myths and presumptions regarding practical aspects of weight loss, only rarely is there sufficient evidence for confirmation or denial [3].

3.4. *David Barker*

Many researchers do not necessarily accept the idea that weight-related non-communicable disease is primarily a behavioural problem. An epidemiologist, David J. Barker, working at the University of Southampton, UK, argued that the best way to explain the onset of non-communicable conditions such as cardiovascular and/or circulatory disease was to postulate an early-age onset. He published his work as "the fetal and infant origins of non-communicable disease" but made his own opinion quite clear with the subtitle "the womb may be more important than the home" [4]. What became known as his "fetal origins hypothesis" continued to fascinate researchers for the next decade and beyond, but without definitive evidence [5], although it continues to make good economic sense [6]. Barker's observations linked what he called "degenerative disease" to poor living conditions in the family: inadequate housing and diet, unemployment and, in his day, large families. There was also a clear inverse relationship between birth weight, blood pressure, and chance of death from ischaemic heart disease and stroke. Finally, Barker linked apparently unrelated conditions such as schizophrenia and obstructive lung disease with a poor maternal environment [4]. Interestingly, he did not add immune system problems to his initial survey of early-onset non-communicable disease, possibly because it was not his field, but also, perhaps, because this aspect was already being covered by his London-based colleague, David Strachan, just a few miles down the road.

3.5. *David Strachan*

Of course, many autoimmune conditions may also be classed as non-communicable. At about the same time as David Barker was developing his foetal origin hypothesis, David Strachan, working in the London School of Hygiene and Tropical Medicine, UK, was linking the nuisance of hay fever with the presence of childhood eczema, and with the greater danger of asthma. He came to the conclusion that being born into larger families provided some protection against atopy, and tentatively suggested that early-life exposure to a more germ-laden environment was the protective factor [7]. This so-called "hygiene hypothesis" struck a chord with the public, and was followed up over many years by Graham Rook and his team, looking for external entities that could, somehow, redirect the immune system to avoid atopic disease. Sadly, no such agent was ever found [8]. Although immune system malfunction comes in a wide variety of disguises, perhaps there are none more surprising than an allergy to utterly harmless grains of pollen.

3.6. *John Bostock*

In the early 19th century, a noted researcher by the name of John Bostock made what he thought must be a significant observation. Born in 1773, in Liverpool, UK, he learnt his trade as a medical doctor at the prestigious medical school in Edinburgh, Scotland, UK, and became famous for studying the function of the kidneys in health and disease. Starting at eight years old, he developed a bizarre condition which, in the fashion of the time, he Latinised as catarrhus aestivus, because it blocked up his nose every summer. Although he tried many remedies, the only one that gave any relief was to leave the town and live by the seaside during the summer months. His description, published in 1819, clearly and unambiguously described what we now know as seasonal allergic rhinitis, or, more commonly, as hay fever [9]. Intrigued, he toured the country, applying strict criteria to check the extent of "his" new disease. His results, published in 1828, were remarkable. Although he took pains to look across all levels of society, he found only 28 cases, all from "the highest ranks of society" [10]. Although he does not specifically say so, it seems likely that these people included the reigning King himself, and it is from this period that the court moved to spend the summer in various seaside towns, the favoured places naming themselves Regis in honour of this fact (and as an opportunity to advertise!). Having identified this new "hay fever" (as the newspapers dubbed it) with royalty, of course everybody wanted it and, accordingly, subsequent epidemiology becomes unclear. Hay fever is exceptional in the sense that people are unsure as to whether such a nuisance is actually a disease, especially if it is quite mild. Nevertheless, the current situation clearly differs from Bostock's day as, more recently, over 40% exhibit the symptoms [11]. Clearly, this vast increase in prevalence of disease must, somehow, be related to its cause. It is worth noting that Bostock himself died of cholera, a common infectious disease of the time, at the very respectable age of 73 years. Although hay fever alone is not a serious condition, nevertheless it indicates that something is going seriously wrong somewhere.

3.7. *Sigmund Freud*

As indicated by Barker, mental health conditions such as schizophrenia also fall under the category of non-communicable diseases. The early practitioner in this very new field was a neurologist working in Vienna in the late nineteenth century called Sigmund Freud [12]. Unfortunately, while hay fever comes with a distinctive set of symptoms, the mental problems treated by his newly developed technique of psychoanalysis defy precise characterisation and, accordingly, we have little idea of the epidemiology of these diseases. Although Freud successfully treated many people "from the highest ranks of society" (the phrase used by John Bostock), unlike with hay fever, sadly, we cannot say with any certainty that the less well-off people did not suffer from these diseases. Unfortunately, for many years after Freud, the "interesting" ideas of the psychoanalysts brought disrepute to the very idea of mental health, especially in the eyes of their fellow medical scientists, a legacy which is only just being overturned.

3.8. Denis Burkitt

As noted in Chapter 2, probably our most important witness to the transition between traditional and modern societies, Denis Burkitt, fell firmly into this era of mental health disbelief. Born in 1911, in Northern Ireland, UK, he qualified as a surgeon in Edinburgh, the same place as did John Bostock, but in the very different context of the diseases of the modern world. Posted to Africa during the Second World War, he returned after that conflict and took his surgery skills across the continent. During this time, Burkitt discovered a virally transmitted sarcoma of the jaw, known nowadays as Burkitt's lymphoma [13]. During his travels, he discovered a more baffling phenomenon that defied explanation according to the knowledge of his day: that there was actually an absence of disease in the traditional societies throughout the bulk of Africa. It would seem that, over the century or so between the hay fever publications of Bostock and the training of Burkitt, the entire panoply of what Burkitt called "Diseases Characteristic of Modern Western Civilization" had become incorporated within the remit of medical instruction (Table 1, Chapter 2) [14]. However, the most striking characteristic of Burkitt's "modern, Western" way of life is the reduction in faecal output to, on average, one-third of the weight (from about 450 g to a much smaller 150 g per person per day), and often a lot less. Accordingly, in a similar fashion to the proponents of lifestyle disease mentioned above, Burkitt specifically blames the "bad habits" of the consumption of refined white flour and the corresponding reduction in dietary fibre, noting specifically that people emigrating from Africa and moving into areas vulnerable to disease tend, over time, to become consistent with the majority population. Burkitt summarised the development of these diseases throughout western European history, noting that their prevalence has increased over the "last 50 years" (i.e., since about 1920), and notes the suggestion from art and literature that obesity only became common after the mid-eighteenth century. Significantly, however, he is at a loss to explain the presence of immune system problems. As discussed above, and in agreement with the mores of the times, he does not mention mental health issues at all [14].

3.9. Adolphe Quetelet

As interest in the obesity phenomenon grew during the 19th century, a mathematician by the name of Adolphe Quetelet developed what became known as the body mass index (BMI—body mass in kilograms divided by the square of the height in metres), a statistical measure which was originally used to compare different populations, but is also convenient for use in individuals [15]. During the Victorian and the following Edwardian times (mid-19th to early 20th century) obesity tended to follow the pattern described by John Bostock and his catarrhus aestivus, in the sense of a patchy distribution among relatively rich people [10]. As an example, a certain William Banting, hailing from a notable family of funeral directors to the Kings of England, and therefore from a well-off background, was born in 1796 and eventually developed a notable paunch. Unlike Bostock's seasonal complaint, which he could only alleviate by living next to the sea throughout the summer, he finally found an acceptable solution in the shape of a low-carbohydrate diet, publishing it in 1863 as a booklet entitled *"Letter on Corpulence, Addressed to the Public"* [16]. Later in the century, Lulu Hunt Peters was born in Maine, USA, in 1873 and, as she termed it, "never grew out of her puppy fat". She eventually reached 100 kg, on the edge of morbidly obese in today's terms. On her graduation as a medical doctor, Peters employed the newly established idea of calories to put herself on a 1200 kcal diet. Upon reaching a more respectable 60 kg, like Banting before her, she realised that there was value in writing a book to reach a wider audience. Published in 1918 under the title *Diet and health: with key to the calories*, she asked her female readers to categorise themselves as either active or inactive, recommending that they eat what she considered to be a relatively low 2250 and 3000 kcal/day, respectively [17].

3.10. Ancel Keys

As the Second World War in Europe drew slowly to its messy conclusion, it was clear that the victors would be faced with the challenge of bringing large numbers of starving people back to health. As it was also known that overfeeding dangerously emaciated people could lead to their death, an experiment was carefully performed on young male conscientious objectors, who wanted to help the war effort but did not feel that they could adopt a role which involved direct combat. Starting in November 1944, the resulting Minnesota Experiment was led by a researcher called Ancel Keys, in which the male volunteers were encouraged to initially adopt a relatively low-calorie "semistarvation" diet for a period of six months, to be subsequently brought back to their previous weights using a variety of methods. Fortunately, Keys' copious notes have recently been summarised in article form, quoting their initial dietary intake as 3200 kcal/day, a number that dropped to 1800 kcal/day in low-calorie mode [18]. Sadly, these volunteers suffered significant psychological trauma, possibly because they were relatively isolated individuals deliberately going hungry in the midst of plentiful, readily available food, unlike situations in which groups of starving people can support one another in a shared attempt to survive the visible absence of food.

3.11. Energy Gain by Food Intake

Needless to say, the estimation of total energy intakes from household purchases can be a complex process, which necessitates accounting for factors such as cooking losses and wastage of inedible materials, as well as meals consumed outside the home or, indeed, fed to pets. A valuable snapshot dates from the year 2007, corresponding to the 40th anniversary of the British Nutrition Foundation, an industry-based advocacy body for health-related policy [19]. According to their data, the estimated energy intake per person peaked in the 1950s at 2660 kcal/person/day which, bearing in mind that these data represent an average of both adults and children, is comparable to Keys' and Peters' adult-only observations of about 3000 kcal/day. Significantly, however, since that time, intake has slumped to a figure of only 1750 kcal/person/day, more comparable to Keys' "semistarvation" figure quoted above, but without the dramatic psychological effects. Interestingly, they made the seemingly definitive, but nevertheless unsupported, statement that this fall is "in line with declining levels of physical activity", echoing the 1912 comments noted above [1], and connecting this with extensive use of the motor car, lifts, television viewing, and computer gaming [19]. In other words, utilising the terminology of Casazza and colleagues, the authors were making a presumption (a view held in the absence of actual evidence) [3].

3.12. Energy Use Estimation by Labelled Water

It seems that the automobile has indeed had a major impact on Burkitt-style "Westernised civilisation", as cars are spreading across the world, roughly in line with the rise in non-communicable disease. Motor cars may be a source of pride to their owners (and envy otherwise), but are they also a source of laziness? Although energy balance determinations can be performed on people living in monitored cages, there is only one method that can be used on free-living individuals: that of double-labelled water (DLW); that is, water in which the normal isotopes of hydrogen-1 and oxygen-16 have been supplemented with so-called labelled water, containing known portions of the heavier isotopes hydrogen-2 and oxygen-18 [20]. Water enters the body either directly as drink or via the oxidation of food, while a portion of it leaves the body in either perspiration or urine. The essence of the method is that, while hydrogen leaves the body essentially exclusively as water, oxygen can leave either in water or in the carbon dioxide produced by physical activity. It follows that the levels of oxygen-18 decline faster than the levels of hydrogen-2, with the rate of such decline over a period of two weeks or so giving a measure of carbon dioxide produced and, hence, energy usage. Perhaps surprisingly, given the common belief that obesity must follow from lower activity levels, they recorded no difference between energy usage in the 1980s and the significantly heavier 2008, when this study was reported. Furthermore, little difference in energy usage was observed between our supposedly lazier world of Westernised civilisation and people from traditional societies. Perhaps most surprisingly, the researchers concluded that humans tend to use the same amount of energy as other mammals of comparable size, implying that our evolved biology is the dominant factor in energy usage [21]. A later review put these observations into a wider context, stating that weight cannot be lost simply by increasing physical activity and concluding, therefore, that a reduction in food intake is the only way to counter the obesity epidemic of the early 21st century [22]. It is interesting to note that this conclusion is exactly the opposite to that published by Foster and Lunn based on data supplied to the British Nutrition Foundation [19]. Indeed, the suspicion is that all these authors were really dealing in Casazza-style myths (views so strongly held that they persist in spite of evidence to the contrary) [3]. Sadly, regardless of all these efforts, it seems likely that the visible dominance of the motor car will continue to be at least partly blamed for the obesity crisis [23]. In reality, however, the influence of modern transport is to allow us to move further and faster than ever before but, when we get there, it seems that we behave in essentially the same way as other mammals of a comparable size [21].

3.13. Business as Usual

Even though the isotope studies reported in the previous paragraph seemed to be definitive, even a decade later, study after study continued to mention behavioural changes. Accordingly, even where the "complex relationship" between the microbiome and the gut–brain axis is treated as an accepted fact, still the rationale for obesity rests on "highly palatable foods coupled with decreased energy expenditure" [24]. A similar article has exploited the BMI-GRS (body mass index–related genetic risk score) to suggest that overeating is brought about by genetically controlled disinhibition, emotional eating, and hunger, implying that ways could be found to encourage restraint [25] but without mentioning the intestinal microbiome. More Casazza-style presumptions! [3] Unfortunately for such arguments, it seems that these epidemics are spreading faster than conventional genetic inheritance mechanisms would allow.

3.14. Weight Gain in Children

The quoted British Nutrition Foundation article mentions "declining levels of physical education and sport in some schools" [19], an observation echoed by a study on the physical fitness of different classes of 10-year-old schoolchildren in England. Carried out over the years 1998, 2008, and 2014, both height and weight had increased proportionately with each generation, but without any alteration in BMI (body mass index). Disconcertingly, however, measures of strength such as sit-ups and handgrip showed a proportionate decline between each generation [26], suggestive, perhaps, of fat penetration between the muscle fibres. As these data were obtained in a sports science context, their (unsurprising) conclusion was that there is a need for a greater emphasis on physical education in school curricula, a presumption which has already been challenged by Casazza, citing a lack of evidence [3]. Similar results have been reported for Slovenian children [27] and for adults undergoing so-called "yoyo" dieting, which tends to lead to an increased proportion of fat [28].

3.15. Cancer Epidemic

As this century of non-communicable disease progresses, we are beginning to see an outbreak of cancer in the younger (below 50s) age group, for which evidence suggests that the cause lies in early life and young adulthood [29]. Similar results have recently been observed for breast cancer in young women (aged 20 to 49 years) [30]. While the immediate cause of cancer may be attributed to exposure to carcinogens [31], it could be that the disturbance of the immune system represented by the hay fever epidemic is actually the primary problem [7]. While a proximate cause for the spread of hay fever may be a change in the nature of pollen with time and, indeed, with anthropogenic climate change [32], there could still be an underlying weakness that led to the immune system being compromised in the first place.

3.16. Mental Health Epidemic

The third major plague of modern times, poor mental health, has risen alongside the expansion of social media, especially in the young, both in Europe [33], and around the world [34]. However, in spite of many studies, a recent systematic review [35] and a network analysis [36] have shown that it is very hard to be sure of a causative connection between social media use and poor mental health. While social isolation is a state of affairs imposed on the individual by circumstances beyond their control, loneliness can be defined as a discrepancy between desired and actual levels of social connectedness that is neither by choice nor specifically imposed [37]. Whatever the underlying causes, loneliness is associated with poor outcomes related to cardiovascular disease, and has been classified as an epidemic, at least in the United States of America [38]. It is worth noting that there are substantial overlaps between poor mental health and immune system conditions; for example, with coeliac disease [39]. Accordingly, a major collaborative investigation has recently been undertaken using the gene-based technique of Mendelian randomisation to investigate the observed links between allergic disease and poor mental health. Although causal links were not supported, strong evidence was found between what they describe as a "broad allergic disease phenotype" and depression [40]. Similar non-causal links were also found with hay fever, asthma, and even Alzheimer's disease, although links to the latter were weaker. The overall conclusion seems to be that a separate factor is causing all these "modern" epidemics, but operating in parallel, rather than sequentially causative [40]. Finally, there have been claims that inflammation is itself contagious through social media-induced behavioural stressors, although they do not seem to include the microbiome within their model [41]. While such "psychopathology" arguments may apply to catastrophising humans, it seems that they should break down if similar problems can be shown to apply in non-human species.

3.17. Animals

Autoimmune-like inflammatory diseases do indeed exist in pet and farmed animals [42], suggesting an interference with biological processes that have been evolving for a long time. While human obesity is blamed on the bad habits of "lifestyle disease", ignoring our overuse of antibiotics [43], antimicrobial use as growth promoters in animal husbandry has been conventionally explained by the reduction in microbial competition for nutrients, thus limiting unspecified maintenance costs on the gastrointestinal system [44]. Instead, it seems likely that the same mechanisms are operating between antibiotic-treated farm animals on the one hand, and humans on the other, both leading to weight gain, autoimmune disease, and poor mental health (the latter in humans at least).

3.18. Summary

As stated above, it seems that the apparently definitive labelled-water experiments concerning energy use were simply ignored. The reason for this was, presumably, due to the absence of a theoretical framework into which the results could be fitted. Accordingly, the stalemate continued: those who studied energy assumed that we eat too much, while those who estimated the amount that we eat assumed that we exercised too little. In the next chapter it is necessary to look more closely at what may be called "post-behavioural" explanations concerning both genes and the bacterial microbiome.

References

1. Jones, D.S.; Podolsky, S.H.; Greene, J.A. The burden of disease and the changing task of medicine. *N. Engl. J. Med.* **2012**, *366*, 2333–2338. [CrossRef] [PubMed]
2. Ling, M.Y.J.; Ahmad, N.; Aizuddin, A.N. Risk perception of non-communicable diseases: A systematic review on its assessment and associated factors. *PLoS ONE* **2023**, *18*, e0286518. [CrossRef]
3. Casazza, K.; Brown, A.; Astrup, A.; Bertz, F.; Baum, C.; Brown, B.B.; Dawson, J.; Durant, N.; Dutton, G.; Fields, D.A.; et al. Weighing the evidence of common beliefs in obesity research. *Crit. Rev. Food Sci. Nutr.* **2015**, *55*, 2014–2053. [CrossRef] [PubMed]
4. Barker, D.J. The fetal and infant origins of adult disease. *BMJ* **1990**, *301*, 1111. [CrossRef] [PubMed]
5. Eriksson, J.G. The fetal origins hypothesis—10 years on. *BMJ* **2005**, *330*, 1096–1097. [CrossRef] [PubMed]
6. Almond, D.; Currie, J. Killing me softly: The fetal origins hypothesis. *J. Econ. Perspect.* **2011**, *25*, 153–172. [CrossRef] [PubMed]
7. Strachan, D.P. Hay fever, hygiene and household size. *BMJ* **1989**, *299*, 1259–1260. [CrossRef] [PubMed]
8. Rook, G.A.W. The old friends hypothesis: Evolution, immunoregulation and essential microbial inputs. *Front. Allergy* **2023**, *4*, 1220481. [CrossRef] [PubMed]
9. Bostock, J. Case of a periodical affection of the eyes and chest. *Med. Chir. Trans.* **1819**, *10*, 161–165. [CrossRef] [PubMed]
10. Bostock, J. Of the catarrhus aestivus or summer catarrh. *Med. Chir. Trans.* **1828**, *14*, 437–446. [CrossRef] [PubMed]
11. Walker, S.; Khan-Wasti, S.; Fletcher, M.; Sheikh, A. Prevalence of hayfever symptoms and diagnosis in UK teenagers. *Prim. Care Respir. J.* **2005**, *14*, 270. [CrossRef]
12. Borch-Jacobsen, M.; Shamdasani, S. *The Freud Files: An Inquiry Into the History of Psychoanalysis*; Cambridge University Press: Cambridge, UK, 2012.
13. Burkitt, D. A sarcoma involving the jaws in African children. *Br. J. Surg.* **1958**, *46*, 218–223. [CrossRef] [PubMed]
14. Burkitt, D.P. Some diseases characteristic of modern western civilization. *Br. Med. J.* **1973**, *1*, 274–278. [CrossRef] [PubMed]
15. Eknoyan, G. Adolphe Quetelet (1796–1874)-the average man and indices of obesity. *Nephrol. Dial. Transplant.* **2008**, *23*, 47–51. [CrossRef] [PubMed]
16. Banting, W. Letter on corpulence, addressed to the public. 1869. *Obes. Res.* **1993**, *1*, 153–163. [CrossRef] [PubMed]
17. Jou, C. The progressive era body project: Calorie-counting and "disciplining the stomach" in 1920s America. *J. Gilded Age Prog. Era* **2019**, *18*, 422–440. [CrossRef]
18. Kalm, L.M.; Semba, R.D. They starved so that others be better fed: Remembering Ancel Keys and the Minnesota experiment. *J. Nutr.* **2005**, *135*, 1347–1352. [CrossRef] [PubMed]
19. Foster, R.; Lunn, J. 40th Anniversary briefing paper: Food availability and our changing diet. *Nutr. Bull.* **2007**, *32*, 187–249. [CrossRef]
20. Schoeller, D.A.; Van Santen, E. Measurement of energy expenditure in humans by doubly labelled water method. *J. Appl. Physiol.* **1982**, *53*, 955–959. [CrossRef] [PubMed]

21. Westerterp, K.R.; Speakman, J.R. Physical activity energy expenditure has not declined since the 1980s and matches energy expenditures of wild mammals. *Int. J. Obes.* **2008**, *32*, 1256–1263. [CrossRef] [PubMed]
22. Westerterp, K.R. Physical activity, food intake, and body weight regulation: Insights from doubly labeled water studies. *Nutr. Rev.* **2010**, *68*, 148–154. [CrossRef] [PubMed]
23. Bonilla, D.A.; Peralta-Alzate, J.O.; Bonilla-Henao, J.A.; Cannataro, R.; Cardozo, L.A.; Vargas-Molina, S.; Stout, J.R.; Kreider, R.B.; Petro, J.L. Insights into non-exercise physical activity on control of body mass: A review with practical applications. *J. Funct. Morphol. Kinesiol.* **2023**, *8*, 44. [CrossRef] [PubMed]
24. Wachsmuth, H.R.; Weninger, S.N.; Duca, F.A. Role of the gut-brain axis in energy and glucose metabolism. *Exp. Mol. Med.* **2022**, *54*, 377–392. [CrossRef] [PubMed]
25. Begum, S.; Hinton, E.C.; Toumpakari, Z.; Frayling, T.M.; Howe, L.; Johnson, L.; Lawrence, N. Mediation and moderation of genetic risk of obesity through eating behaviours in two UK cohorts. *Int. J. Epidemiol.* **2023**, *52*, 1926–1938. [CrossRef] [PubMed]
26. Sandercock, G.R.H.; Cohen, D.D. Temporal trends in muscular fitness of English 10-year-olds 1998–2014: An allometric approach. *J. Sci. Med. Sport* **2019**, *22*, 201–205. [CrossRef] [PubMed]
27. Đuric, S.; Sember, V.; Starc, G.; Soric, M.; Kovac, M.; Jurak, G. Secular trends in muscular fitness from 1983 to 2014 among Slovenian children and adolescents. *Scand. J. Med. Sci. Sports* **2021**, *31*, 1853–1861. [CrossRef] [PubMed]
28. Yates, T.; Biddle, G.J.H.; Henson, J.; Edwardson, C.L.; Arsenyadis, F.; Goff, L.M.; Papamargaritis, D.; Webb, D.R.; Khunti, K.; Davies, M.J. Impact of weight loss and weight gain trajectories on body composition in a population at high risk of type 2 diabetes: A prospective cohort analysis. *Diabetes Obes. Metab.* **2024**, *26*, 1008–1015. [CrossRef] [PubMed]
29. Ugai, T.; Sasamoto, N.; Lee, H.Y.; Ando, M.; Song, M.; Tamimi, R.M.; Kawachi, I.; Campbell, P.T.; Giovannucci, E.L.; Weiderpass, E.; et al. Is early-onset cancer an emerging global epidemic? Current evidence and future implications. *Nat. Rev. Clin. Oncol.* **2022**, *19*, 656–673. [CrossRef] [PubMed]
30. Xu, S.; Murtagh, S.; Han, Y.; Wan, F.; Toriola, A.T. Breast cancer incidence among US women aged 20 to 49 years by race, stage, and hormone receptor status. *JAMA Netw. Open* **2024**, *7*, e2353331. [CrossRef] [PubMed]
31. Blackadar, C.B. Historical review of the causes of cancer. *World J. Clin. Oncol.* **2016**, *7*, 54–86. [CrossRef] [PubMed]
32. D'Amato, G.; Murietta-Aguttes, M.; D'Amato, M.; Ansotegui, I.J. Pollen respiratory allergy: Is it really seasonal? *World Allergy Organ J.* **2023**, *16*, 100799. [CrossRef] [PubMed]
33. Sacco, R.; Camilleri, N.; Eberhardt, J.; Umla-Runge, K.; Newbury-Birch, D. A systematic review and meta-analysis on the prevalence of mental disorders among children and adolescents in Europe. *Eur. Child Adolesc. Psychiatry* **2022**, *33*, 2877–2894. [CrossRef] [PubMed]
34. Steel, Z.; Mamane, C.; Iranpour, C.; Chey, T.; Jackson, J.W.; Patel, V.; Silove, D. The global prevalence of common mental disorders: A systematic review and meta-analysis 1980–2013. *Int. J. Epidemiol.* **2014**, *43*, 476–493. [CrossRef] [PubMed]

35. Keles, B.; McCrae, N.; Grealish, A. A systematic review: The influence of social media on depression, anxiety and psychological distress in adolescents. *Int. J. Adolesc. Youth* **2020**, *25*, 79–93. [CrossRef]
36. Panayiotou, M.; Black, L.; Carmichael-Murphy, P.; Qualter, P.; Humphrey, N. Time spent on social media among the least influential factors in adolescent mental health: Preliminary results from a panel network analysis. *Nat. Ment. Health* **2023**, *1*, 316–326. [CrossRef]
37. Surkalim, D.L.; Luo, M.; Eres, R.; Gebel, K.; van Buskirk, J.; Bauman, A.; Ding, D. The prevalence of loneliness across 113 countries: Systematic review and meta-analysis. *BMJ* **2022**, *376*, e067068. [CrossRef] [PubMed]
38. Tanne, J.H. Epidemic of loneliness threatens public health, says US surgeon general. *BMJ* **2023**, *381*, p1017. [CrossRef] [PubMed]
39. Zingone, F.; Swift, G.L.; Card, T.R.; Sanders, D.S.; Ludvigsson, J.F.; Bai, J.C. Psychological morbidity of celiac disease: A review of the literature. *United Eur. Gastroenterol. J.* **2015**, *3*, 136–145. [CrossRef] [PubMed]
40. Budu-Aggrey, A.; Joyce, S.; Davies, N.M.; Paternoster, L.; Munafò, M.R.; Brown, S.J.; Evans, J.; Sallis, H.M. Investigating the causal relationship between allergic disease and mental health. *Clin. Exp. Allergy* **2021**, *51*, 1449–1458. [CrossRef] [PubMed]
41. Vodovotz, Y.; Arciero, J.; Verschure, F.M.J.; Katz, D.L. A multiscale inflammatory map: Linking individual stress to societal dysfunction. *Front. Sci.* **2024**, *1*, 1239462. [CrossRef] [PubMed]
42. Marsella, R. Atopic dermatitis in domestic animals: What our current understanding is and how this applies to clinical practice. *Vet. Sci.* **2021**, *8*, 124. [CrossRef] [PubMed]
43. Shapiro, D.J.; Hicks, L.A.; Pavia, A.T.; Hersh, A.L. Antibiotic prescribing for adults in ambulatory care in the USA, 2007–09. *J. Antimicrob. Chemother.* **2014**, *69*, 234–240. [CrossRef] [PubMed]
44. Gaskins, H.R.; Collier, C.T.; Anderson, D.B. Antibiotics as growth promotants: Mode of action. *Anim. Biotechnol.* **2002**, *13*, 29–42. [CrossRef] [PubMed]

4. The Cause of Disease: Genetic/Infectious/Dietary Hypotheses

4.1. The Triple Plagues

While the initial focus on weight gain resulting from eating too much and/or moving too little remains superficially attractive, it is clear that the underlying facts of weight gain do not support the argument, and that there are parallel "plagues" of rising immune system and mental health diseases to be taken into account. Nevertheless, by and large, the "gene-inheritance" hypothesis still dominates early 21st century discussion, i.e., something unknown, be it infection or deficiency, has imposed itself on our basic gene-defined humanity. In turn, the microbiome, if mentioned at all, is looked on as both unknown and, probably, irrelevant to the overall discussion. For example, one recent review of the immune system, while mentioning the microbiome, nevertheless comes to no specific conclusion about its relevance [1]. However, on the whole, the concept of the microbiome consistently, insistently intrudes itself into the more recent discussions, albeit with too great an emphasis on the presence of the more easily detected bacteria and fungi and with relatively little effort put into searching for potential beneficial microeukaryotes.

4.2. Life-Forms

The early philosopher–scientists were faced with many problems, amongst which were naming the variety of plants and animals that they saw around them. Carl Linnaeus helped when he introduced his system of binomial nomenclature: generic plus trivial, which is fine for the multicellular eukaryotes such as *Panthera Leo*, for example, with his shaggy mane and sharp teeth. However, looking more carefully illustrates the need for a third layer of distinction [2], such as to distinguish between different isolated populations of species such as the western lowland gorilla: *Gorilla gorilla gorilla*. However, in a dilemma that has been called "the species problem" [3], we must ask the following question: in what way are such distinctions actually relevant? One recent suggestion has been to eliminate all species-level names and to count only the "SNaRC", the Smallest Named and Registered Clade, claiming that this would solve all the "theoretical tasks of biology, in evolution and ecology" [4]. These tasks continue to astonish, as a recent observation of the "horizontal" uptake of numerous plant genes into the herbivorous whitefly *Bemisia tabaci* illustrates [5]. Since the Linnaean system is so widely used, perhaps it is not surprising that it is also applied to microbes such as the bacterium *Streptococcus pneumoniae*. In an article published in 1928, however, a researcher by the name of Frederick Griffith described experiments on two strains of the diploid version of this bacterium, one lethal to mice, one harmless. By degrading the lethal version, and stirring the harmless one with the resultant innocuous debris, he showed that this previously harmless strain had been "upgraded" to become capable of killing mice [6]. While Griffith called this phenomenon a "transforming principle," and it helped to initiate the genetics revolution, the ability to cause death is clearly relevant to the nature of individual microbial "species," but it is an ability that current systems of nomenclature seem unable to completely capture. Many questions are left unanswered. For example, is a microbial community such as a microbiome a living entity in its own right [7]? Or does it resemble a biofilm, which can be defined by its physical traits [8], and which can include more than only bacteria [9]?

4.3. *Ubiquitous Bacteria*

A recent set of observations has raised yet further questions—is the presence of bacteria in the brain of recently deceased people a cause of their disease, a harmless commensal, a contaminant, or is it a sign that the brain itself has its own microbial community, a supportive mutualistic microbiome [10]? Equally, does the apparent discovery of bacteria inside the theoretically sterile human placenta provide evidence of a pre-birth microbiome, an infective cause of infant-onset non-communicable disease, or simply contamination [11]? In a sense, it could be that the developmental problems induced by a malfunctioning microbiome are analogous to the deficiency due to the inheritance of recessive genes, potentially illustrated by the example of the so-called "Habsburg jaw" [12]. Indeed, it may be argued that there are two primary kinds of microbe-related disease causation: microbiome deficiency on the one hand, and pathogen infection on the other.

4.4. *Infection*

Probably the most perplexing story of recent times was the seemingly sudden appearance of bovine spongiform encephalopathy, "mad cow disease," that burst onto the world stage in the late 20th century. This was an example of what became known as a transmissible spongiform encephalopathy (TSE) and turned out to be a protein containing no nucleic acids at all, a fact which has illuminated hitherto unknown areas of biology [13]. The causative agent became known as a prion, standing for "proteinaceous infectious particle," while the system that it affects, uniquely, is named after the agent itself, as the "major prion protein". Although the overall story is only slowly coming into focus, it seems that this original protein relies on its flexibility to perform its function, while the infectious protein locks it into only one form, a defect that slowly spreads throughout the brain. Studies continue to this day [14], including a new protein-based origin-of-life hypothesis [15]. Clearly, it is necessary to consider all the options when faced with a new class of disease. Although as yet there is no suggestion of a role for microbes in TSE pathology, there would undoubtably be an evolutionary logic to the existence of major prion proteins and it would be foolish, at this early stage, to rule out microbial involvement completely. Indeed, perhaps the single defining feature of microbiome-related disease is the steady rise in its incidence in recent decades. Perhaps coincidentally, it is worth noting that there is also an ongoing rise in cases of the various transmissible spongiform encephalopathies [16]. In addition, very rare cases of Alzheimer's disease may have been transferred along with cadaver-derived pituitary growth hormone. The suggestion, which is still controversial, derives from the observation of subclinical amyloid-beta in patients who died relatively early from medically derived prion-related Creutzfeldt–Jakob disease [17]. Interestingly, Alzheimer's disease has indeed been linked to microbiome malfunction [18], but the full extent of any links between infectious prions, amyloid-beta, and microbiota remains to be elucidated.

4.5. *Oral Bacteria*

Although the idea of "good bacteria" has become established in recent years, there is still scope for old-fashioned infection by either good, neutral, or pathogenic bacteria [19]. Of course, what has been described as the oral microbiome [20] affords access to both the gut–brain axis, for example, with Parkinson's disease [21], and to the so-called gut-lung axis [22], making the mouth the key portal to the body. Accordingly, the immune system is critical, both for the infant oral microbiome [23] and the collection of lymphoid tissue called Waldeyer's ring, including the tonsils and adenoids [24].

4.6. *The Leaky Gut*

It is necessary to keep an open mind when faced with apparently new diseases—are they solely due to infectious agents, to some kind of deficiency or, indeed, an infectious agent exploiting/interacting with a deficiency? In fact, the microbe-related words commensal, parasite, and pathogen likely represent a continuum between opportunity and threat. Probably the best example of this is seen in the elusive set of conditions expressed by the term "leaky gut", a degree of gut barrier disruption known to be affected by, and affecting, many seemingly different conditions [25]. In a similar manner, a recent study has identified some of the factors behind the protection afforded against infection by a diverse microbiome. The authors suggest that nutrient availability is the key to successful pathogen establishment, and high microbiome diversity effectively closes this opportunity [26].

4.7. *Laboratory Microbiomes*

Of course, the predominant lesson taught throughout the 20th century is that germs cause disease. Accordingly, the 2006 discovery (in mice) of a transmissible "obesity-associated gut microbiome with increased capacity for energy harvest" [27] was initially treated with a mixture of excitement and relief. At last, obesity may be cured by antibiotics! In reality, this was just another stage in the slow untangling of the complex interplay between microbe and mouse, let alone between microbiome and man. Another instalment in this long-running series was reported in the journal *Nature* in 2020. While performing experiments related to Lou Gehrig's disease (amyotrophic lateral sclerosis), the researchers found that one mouse model gave significantly different results from genetically identical mice raised by a different supplier. Tracking down the problem by means of faecal transplantation, they identified a strain of bacteria that interacted with the immune system, present in the mice supplied by the first company, but absent in the second [28]. In light of Griffith's previously mentioned experiment [6], there is the possibility of horizontal gene transfer to take into consideration. In his case, further experiments showed that the "transforming principle" was mediated by nucleic acids rather than proteins, kickstarting the DNA revolution but, perhaps, the "transformation" itself was not fully appreciated at the time. Although now known as horizontal gene transfer, this mechanism is responsible for the spread of antibiotic genes into, and out of, environment, clinic, and patient [29]. It could be that the high microbial diversity of the fully functioning microbiome is needed because it can express many key mobile genetic elements. Accordingly, both people living in an industrialised environment, and laboratory-raised animals, many suffer from variations of dysbiosis, perhaps randomised for humans but regimented in the laboratory [30]. No matter how genetically identical the individuals are, in the light of microbiome-function deficiency disease, how can we be sure that animal models are actually useful?

4.8. Missing Microbiota

Initiated by Martin Blaser, what has become known as Blaser's hypothesis states that many of the non-communicable diseases of what he calls "post-modern" society may be caused by the antibiotic-led removal of the key bacteria with which we co-evolved [31]. Thus, while it was once thought that various strains of the bacterium *Helicobacter pylori* must be removed to cure stomach ulcers, it now appears that the use of antibiotics to ensure their complete removal may carry a sting in the tail [32]. In a thorough assessment, while acknowledging the potential of *H. pylori* removal for treatment within the context of various cancers, attempting to cure gastroesophageal reflux disease may actually worsen the situation, presumably by provoking dysbiosis-like disturbance of the resident microbes [32]. The condition most closely associated with antibiotic overuse, however, is *C. difficile* (*Clostridioides difficile*, often known by its earlier name of *Clostridium difficile*) overgrowth [33], a condition often reported to be dramatically improved by faecal microbiota transplantation (FMT), although this procedure is not always successful under the stricter conditions of randomised clinical trials [34].

4.9. Probiotics

In the context of Blaser's hypothesis, the idea that missing microbes can be replaced and/or supplemented seems to be intuitively correct. In principle, a single, precisely defined bacterium, for example, could be developed and licenced as a valid pharmaceutical agent, hopefully curing disease and affording healthy profit margins [35]. Although initial signs were promising, sadly, recent developments in the field seem to be running into the same problems faced by the wider pharmaceutical industry: both inconclusive results, and persistent questions about the exact mechanism of action. Examples include the not entirely successful search for a probiotic to act against irritable bowel syndrome and their failure to bring about "persistent changes in gut microbiota" [36], to improve the lipid profile in the blood of otherwise seemingly healthy adults [37], or to convince everybody as whether they are, indeed, safe and effective [38]. In essence, we have a similar situation to the ongoing search for infectious agents as a cause of non-communicable disease, namely a potentially infinite search for the single, well-defined, patentable entity that will act as a cure. Owing to the uncertainty pervading the field, Harald Brüssow has called for a germ theory-like effort to uncover the link between specific microbes and corresponding disease states, and also for a deeper analysis of the ecological and evolutionary logic surrounding probiotics [39]. As a case in point, as a part of the ongoing search for a probiotic that may prevent necrotising enterocolitis in premature babies, researchers alighted on a choice between two *Clostridium* (*Clostridioides*) bacterial species, *butyricum* or *tyrobutyricum*, deciding, on the basis of preliminary work in mice, that the former is a "devil", i.e., a pathogen, and the latter an "angel", a (potential) probiotic [40]. As illustrated by Griffith's experiment discussed above [6], there is a deeper problem here, a question posed by the flexibility of horizontal gene transfer.

4.10. *Jekyll or Hyde?*

Both probiotics and pathogens are external microbial agents interacting with the immune system and the pre-existing microbiome. In the light of the problems noted above regarding the reproducibility of experiments involving laboratory animals [28], and the renewed interest in horizontal gene transfer, it seems that our understanding of the microbiome itself is incomplete. It may be that, rather than angels and devils [40], probiotics and pathogens are more like Jekyll and Hyde (from the story by Robert Louis Stevenson, *The Strange Case of Dr Jekyll and Mr Hyde*, in which a "serum" converts the good-natured Dr. Jekyll into the murderous Mr. Hyde). In this context, an individual microbial species may display very different activity when expressing genes transferred "horizontally" from the microbiome, an action analogous to Griffith's transforming principle, in which *S. pneumoniae* may be converted from harmless to lethal under certain circumstances [6]. Griffith was working according to the precepts of the germ theory of disease, which followed from Pasteur's fermentation studies conducted at least half a century earlier. While it might be thought that modern genomic studies might help to detect causative agents, a short communication on the connection between acne vulgaris and its eponymous associated bacterium, *Propionibacterium acnes*, illustrates that this is not the case [41]. Rather than looking at the individual bacterial "angel or devil" constituents, it seems to be necessary to consider the "Jekyll or Hyde" characteristics of the microbiome as a whole.

4.11. *Solely Genetic Inheritance?*

By the late 20th century, it was clear that, for whatever reason, dietary fibre alone was not going to provide the answer to the riddle of obesity. By contrast, the sequencing of the human genome was being considered and, accordingly, a gene-based obesity study was envisioned by Claude Bouchard and colleagues at the University of Quebec in Canada. This was performed by overfeeding 12 pairs of young, genetically identical (monozygotic) male twins for 7 weeks after an initial 2-week observation period. In principle, if weight gain were solely under the control of genes inherited from either parent, then any variance would be lower (i.e., more similar) within such pairs than between them, and this discrepancy was indeed observed [42]. Although the reported conclusion of this study was that both energy retention and the distribution of excess fat around the body were under the control of "undetermined genetic characteristics", both the wider spread of obesity and the size of the inter-pair differences in this experiment are, perhaps, hard to account for by genetic mechanisms alone. By contrast, the relatively small intra-pair variability would be a good fit for something of variable quality, "Jekyll or Hyde", passed on from either parent. Unfortunately, the birth circumstances of these twin pairs were not recorded. Granted that they were born in the 1950s and 1960s, however, it is likely that they were born by natural methods and that it is the variable maternal microbiome that accounts for the findings of this study. If a similar study were to be repeated in the 2000s, when caesarean section delivery of twins was much more likely, an inter- and intra-pair variability of comparably large magnitude might have been expected. Sadly, perhaps, such deliberate overfeeding would not be ethically acceptable nowadays. Classically, twin studies were used to determine the percentage genetic heritability of different traits. The basic idea is that, genetically identical or not, having the same birth mother and similar upbringing should account for environmental effects, leaving any differences due solely to genetic inheritance. By this means, height was determined to be about 80% heritable [43], although a later search for the genes involved in determining height uncovered hundreds or, indeed, thousands [44]. It could be that there is something seriously wrong with this concept of (solely) genetic inheritance.

4.12. Missing Heritability

In the early 1960s, when the epidemic of obesity was beginning to perplex the genetic epidemiologists, Professor James Neel was attempting to explain why type 2 diabetes was becoming so common. As it showed a strong tendency to run in families, they considered it to have a strong genetic component. Accordingly, Neel put forward the tentative argument that, while some groups of people had evolved so-called "thrifty" genes to be of value in times of dearth, such genes encourage the accumulation of fat, and are therefore positively harmful in times of plenty [45]. Although Neel's hypothesis has remained in limbo, neither confirmed nor denied, the increasing understanding of the different varieties of type 2 diabetes, along with the recent failure to find more than a trace of suspect genes [46], eventually made this approach untenable. In a similar fashion, many 20th century population studies had seemingly established large heritabilities for many common traits. However, as the human genome has slowly been uncovered, it has become clear that these diseases do not have a strong genetic component. Perhaps the most extreme position is that these pre-genomic population studies were simply irrelevant [47] and, at least as far as psychiatry is concerned, this new mood confirming the uselessness of genetic studies is slowly becoming the norm [48]. As an alternative to a genetic problem is an environmental concern, attention has been shifting to potentially microbe-related recent changes in our diet.

4.13. Dietary Deficiency?

As exclusively gene-based explanations for modern "Westernised" disease are beginning to seem less likely, another hypothesis has been taking shape: that the failure of the intestinal microbiome limits the supply of valuable chemicals, including vitamins and microbe-produced SCFAs (short-chain fatty acids). The basic principle is that these SCFAs, the two-carbon-atom acetate (ethanoate), three-carbon propionate (propanoate), and four-carbon butyrate (butanoate), among others, fuel the cells lining the wall of the intestine. Accordingly, any shortage is considered to have knock-on effects around the body [49], with the resultant metabolic deficiency not only causing obesity via the gut–brain axis [50] but also degrading the immune system [51] and affecting the brain itself [52], possibly alongside a link to serotonin production in the enterochromaffin cells lining the gut [53]. The connection between SCFAs and probiotics has stimulated much investigation [54], possibly in combination with prebiotics [55], and studies often, but not always, report positive benefits [36]. Alongside these studies, an influential experiment has been carried out in which microbial extinction has been demonstrated to occur over many generations of mice raised with limited diets [56], and we have seen that key microbes can be lost when mice are brought up under laboratory conditions, presumably involving a restricted diet [28]. In turn, this dietary deficiency hypothesis has been associated with Burkitt's suggestion of an "environmental" problem brought about by low-fibre diets [57]. Interestingly, however, while Burkitt espoused the fibre hypothesis, he also noted that one of the East African peoples that he observed, the Maasai (Masai in his day), remained healthy with relatively low levels of dietary fibre, somewhat undermining his theories [58].

4.14. The Heart-Healthy Maasai

Burkitt states that ischaemic heart disease was a common cause of death in the high-income Westernised societies of his day, but was almost unknown in the traditional peoples that he studied. Significantly, however, the steppe-dwelling, cattle-rearing Maasai (alternative spelling Masai) did not eat the high-fibre diets commonly encountered in the more diversely vegetated areas of the continent, and yet remained essentially free from arterial heart disease. Burkitt's explanation was that, exceptionally, the Maasai had evolved genes with a unique capability of rendering dietary cholesterol harmless; however, he then goes on to admit that "many other factors almost certainly contribute" and "that a fibre-depleted diet may be an important factor which has been overlooked", suggesting that he was not very confident in this explanation. Sadly, however, the normally indefatigable Burkitt did not search for evidence of such people in his contemporary "Westernised" world [58]. One of the more closely studied peoples on the African continent are the hunter–gatherer Hadza of Tanzania. Interestingly, an early microbiome—actually bacteriome—study reported the absence of *Bifidobacteria* in the Hadza [59], which is perhaps surprising, as it is normally classed as a "good" bacterium in a Westernised context [36]. A later study used ultra-deep microbiome genome sequencing to uncover Hadza microbes that are said to be vanishing from modern environments [60]. Basically the same process is playing out in the Amazon regions, with a gain in weight in people from the Venezuelan Amazon being documented as they moved into cities [61], albeit without any mention of their microbiomes. By contrast, an isolated village of Yanomami Amerindian people were reported to possess high bacterial diversity and, interestingly, significant levels of genes coding for resistance to antibacterial agents [62]. Alongside these investigations, changes in the microbiome of the Amazonian peoples are being documented as they become progressively more urbanised, and the term "tropical urban" is being employed to describe ongoing worldwide changes [63]. Likewise, the Tsimane are a people living in the Bolivian Amazon, pursuing a mixed farmer–hunter–gatherer existence, who have been under medical care since 2012. Although it is not entirely clear from such factors as cigarette smoking and antibiotic use, and "suffering" from high systemic inflammation (normally considered a diseased state), they have perhaps the lowest level of coronary atherosclerosis known [64] and a "relatively modest" decrease in brain volume with age [65]. Equally, age has little effect on arterial calcification and blood pressure [66]. A recent study followed the distribution of microbiota among mother–infant pairs, emphasising the modernisation of lifestyle and the correspondingly rapid narrowing of opportunities to learn about their resilience to non-communicable disease [67].

4.15. *Venus Figurines*

No survey of obesity and its causes would be complete without a reference to the mysterious Venus figurines discovered at Upper Palaeolithic sites in Europe [68]. The fact that these clearly obese ladies (whether real or not) were represented in stone suggests that they were significant in some way, and it could be that they represent real women exposed to the same factors that predispose modern-day people to obesity. If so, sadly, we cannot tell if they considered their condition to be a blessing or a curse. While we have no easy access to stone age microbiomes, palaeofaecal samples of up to 2000 years old were obtained from the southwestern USA and Mexico, assembling an approximate diversification timeline for the archaeal species *Methanobrevibacter smithii* and observing an "enrichment of mobile genetic elements" relative to industrialised gut microbiomes [69]. Granted that the current emphasis is firmly on the quality of the "Westernised" food that we eat, how does our food compare with earlier times?

4.16. *Nouvelle Cuisine?*

Even though our deep-time ancestors would not recognise many of the foodstuffs available in modern days, the main differences relate to the extensive processing and increased palatability of food in the present day, while bearing in mind that small amounts of assorted additives may have profound effects. Indeed, the designation "ultraprocessed" can now be used as shorthand for an unhealthy diet, however constituted [70]. Sadly, the exact mechanism by which such diets degrade our health is not entirely clear, although they have been associated with addiction [71] and with depression, with at least one report implicating the use of novel sweeteners [72]. Hyper-palatability is often associated with the hyperphagia (i.e., greed) and reduced physical activity (i.e., laziness) of lifestyle disease, more technically described as an out-of-kilter "energy balance model" (EBM). This EBM concept has been reviewed by Hall et al. [73], but its unsatisfactory nature has driven an alternative model in which the speed of carbohydrate uptake disrupts normal insulin behaviour, leading to the over-accumulation of fat. This "carbohydrate–insulin model" (CIM), championed by David Ludwig and his team, has been summarised [74], but also disagreed with, if only on the basis of a lack of good medical evidence [75]. While investigations continue [76], perhaps there is better evidence to say that improved health outcomes tend to follow from the consumption of complex "high-quality" carbohydrates with a low glycaemic index [77], more reminiscent of Denis Burkitt's dietary fibre [58].

4.17. *Fancy Fats?*

Significantly, perhaps, while the carbohydrate fraction of the diet has changed over the centuries, fatty acid composition has also changed, if anything even more dramatically [78]. The essence of this argument is that our biology has evolved to handle an approximately equal mixture of omega-6 and omega-3 fatty acids, with an even greater proportion of omega-3 where fish was a major component of the diet. However, following the discovery that seed oil-based foodstuffs fattened up food animals much more effectively than (omega-3-containing) grass feeding, the estimated proportion of omega-6 intake has increased at least tenfold in high-income countries, suggesting a strong link with obesity in humans [78]. By contrast, although there is theoretical evidence that a diet high in omega-6 fatty acids ought to lead to high levels of inflammation, these theories do not seem to be backed up by either experiment or epidemiology, leaving the influence of diet essentially unresolved [79]. Similarly, while efforts are being directed toward a dietary inflammatory index, which does not seem to include any information on the microbiome, progress is slow [80]. While commenting upon the recent success of the anti-obesity GLP-1 agonists, even without precise knowledge of the reasons that people gain weight in the first place, Dariush Mozaffarian expresses the current view of "complex interrelated biological interactions" including food processing and "intergenerational transmission of risk" [81]. In this latter category, he includes "epigenetics, noncoding RNAs, microbial species" which, somehow, promote "escalating obesity, even with stable energy intake". Finally, he makes an impassioned plea for urgent investment in research so as to rigorously evaluate scientific hypotheses [81]. Note, however, that while the primary focus has been on energy intake, perhaps we should be looking in exactly the opposite direction.

4.18. *Intestinal Gymnastics*

The most striking characteristic of the lifestyle contrasts described by Denis Burkitt is that the weight of faecal output of people living in the traditional societies that he studied was, on average, three times the offering of people living in what he called "Modern Western Civilization" (i.e., about 450 g/person/day compared to a more "modern" 150 g/person/day) [58]. In a similar fashion, Burkitt describes a much faster intestinal transit time of approximately 35 h in such people, compared to the 70 h reported by Burkitt for "modern" English volunteers [58]. As illustrated in Figure 4, data from more recent times are in broad agreement with Burkitt's findings, with high-income countries producing about the same weight of untreated stool per capita, at 130 g, while low-income countries afford a more substantial 260 g, about halfway towards the output reported for peoples living in wholly traditional societies [82]. While Burkitt suggested that a greater input of dietary fibre might give rise to these results, and the association of low faecal output with the non-communicable diseases of Table 1 (Chapter 2), it is interesting to note the above-mentioned absence of such diseases in the Maasai, who did not consume large amounts of dietary fibre [58].

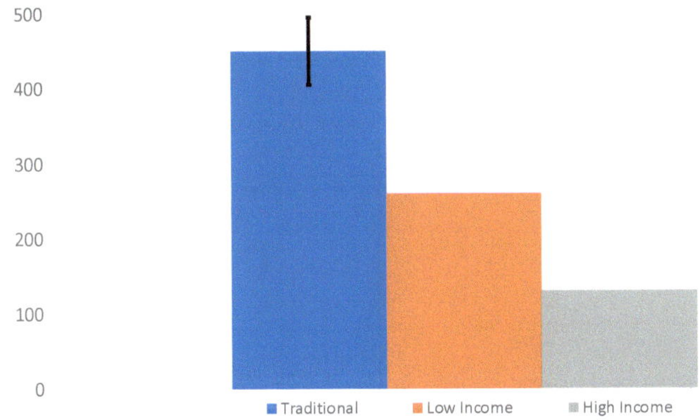

Figure 4. Faecal weight versus type of society. The columns on the left and right correspond to the faecal weights of people living in Burkitt's "traditional" and "Modern, Western Civilization", respectively, with the range bar on the left-hand column corresponding to the range of Burkitt's figures [58]. The centre column represents the data provided by Rose [82] (income levels according to the Human Development Index), effectively an average of the other two. Source: Author's compilation based on data from Burkitt [58] and Rose [82].

4.19. Faecal Energy Excretion

Of course, the increasing tendency toward obesity is the most striking characteristic of Burkitt's "Modern, Western" diseases, both in his day and in the present century [58]. While this is usually ascribed to the overconsumption of hyperpalatable foods, his observations point to exactly the opposite reason: that weight gain is related to an overall reduction in excretion. Interestingly, while Burkitt made sure that his theories were widely known and has probably changed the dietary habits of millions of people over the years, obesity rates have kept on rising and, although hard to define precisely, the levels of chronic constipation have not decreased over the decades [83]. Burkitt also mentions stool quality, which was "soft, unformed" in people living their traditional lifestyle compared to "hard, formed" in modern, Westernised societies. Indeed, it seems that this characteristic is the key, with softer stool consistency being associated with high bacterial diversity and growth rates [84]. The overall picture seems clear: something in the transition from "traditional" to "modern, Western" has decreased bacterial diversity and growth rates, leading to hard faecal consistency and overall accumulation of body weight. Sadly, it seems that excess, unwanted fat is not so easily removed.

4.20. Defending the Fat

Set-point theory states that, along with vital characteristics like water content and body temperature, body weight is detected and maintained according to a value set by the nature of the species [85]. Energy compensation (also known as adaptive homeostasis) is one of the mechanisms used for the defence of body weight set-an effect that has been comprehensively re-evaluated by Careau, Halsey, and others [86]. These workers note that energy compensation results from reduction in basal energy expenditure and, importantly, state that the "degree of energy compensation varied between people of different body composition". In essence, individuals bearing a greater body weight than is appropriate for their species are engaging in permanent exercise, requiring permanent compensation, as every effort is carrying an extra burden. Interestingly, it is possible that the observed drop in average body temperature of US army personnel noted between the mid-19th century and the present day [87] may well be due to the "compensatory" reduction in non-exercise activity thermogenesis [88], a form of basal energy expenditure, following the increase in average body weight over these decades.

4.21. Summary

Although problems relating to mental health may be essentially invisible, even to the sufferers themselves, and immune conditions may be hidden from the world at large, obesity is visible to all. Accordingly, much effort has focused either on a "thrifty" genetic cause, excess of simple carbohydrates or the wrong kind of fat, or hyper-palatability leading to overconsumption (greed), allied with a lack of movement (laziness). As detailed above, however, all these explanations have their problems. While these conditions are associated with industrialisation, the consequences of pollution have not been rigorously followed up. Equally, there has been too great an emphasis on the detection of bacteria, to the detriment of the more complex eukaryotic "benegens" that may be more valuable. As indicated earlier, bacterial dysbiosis is the "smoking gun". Where is the bullet?

References

1. Liston, A.; Humblet-Baron, S.; Duffy, D.; Goris, A. Human immune diversity: From evolution to modernity. *Nat. Immunol.* **2021**, *22*, 1479–1489. [CrossRef] [PubMed]
2. Winston, J.E. Twenty-first century biological nomenclature-the enduring power of names. *Integr. Comp. Biol.* **2018**, *58*, 1122–1131. [CrossRef] [PubMed]
3. Hey, J. The mind of the species problem. *Trends Ecol. Evol.* **2001**, *16*, 326–329. [CrossRef] [PubMed]
4. Mishler, B.D.; Wilkins, J.S. The hunting of the SNaRC: A snarky solution to the species problem. *Philos. Theory Pract. Biol.* **2018**, *10*, 001. [CrossRef]
5. Gilbert, C.; Maumus, F. Multiple horizontal acquisitions of plant genes in the whitefly *Bemisia tabaci*. *Genome Biol. Evol.* **2022**, *14*, evac141. [CrossRef] [PubMed]
6. Griffith, F. The significance of pneumococcal types. *J. Hyg.* **1928**, *27*, 113–159. [CrossRef] [PubMed]
7. Berg, G.; Rybakova, D.; Fischer, D.; Cernava, T.; Vergès, M.-C.C.; Charles, T.; Chen, X.; Cocolin, L.; Eversole, K.; Corral, G.H.; et al. Microbiome definition re-visited: Old concepts and new challenges. *Microbiome* **2020**, *8*, 103. [PubMed]

8. Lichtenberg, M.; Coenye, T.; Parsek, M.R.; Bjarnsholt, T.; Jakobsen, T.H. What's in a name? Characteristics of clinical biofilms. *FEMS Microbiol. Rev.* **2023**, *47*, fuad050. [CrossRef] [PubMed]
9. Flemming, H.C.; Wuertz, S. Bacteria and Archaea on earth and their abundance in biofilms. *Nat. Rev. Microbiol.* **2019**, *17*, 247–260. [CrossRef] [PubMed]
10. Link, C.D. Is there a brain microbiome? *Neurosci. Insights* **2021**, *16*, 26331055211018709. [CrossRef] [PubMed]
11. Panzer, J.J.; Romero, R.; Greenberg, J.M.; Winters, A.D.; Galaz, J.; Gomez-Lopez, N.; Theis, K.R. Is there a placental microbiome? A critical review and re-analysis of published placental microbiota datasets. *BMC Microbiol.* **2023**, *23*, 76. [CrossRef] [PubMed]
12. Vilas, R.; Ceballos, F.C.; Al-Soufi, L.; González-Garcia, R.; Moreno, C.; Moreno, M.; Villanueva, L.; Ruiz, L.; Mateos, J.; González, D.; et al. Is the "Habsburg jaw" related to inbreeding? *Ann. Hum. Biol.* **2019**, *46*, 553–561. [CrossRef] [PubMed]
13. Asher, D.M.; Gregori, L. Chapter 1—Human transmissible spongiform encephalopathies: Historic view. *Handb. Clin. Neurol.* **2018**, *153*, 1–17. [PubMed]
14. Kovač, V.; Šerbec, V.Č. Prion protein: The molecule of many forms and faces. *Int. J. Mol. Sci.* **2022**, *23*, 1232. [CrossRef] [PubMed]
15. Jheeta, S.; Chatzitheodoridis, E.; Devine, K.; Block, J. The way forward for the origin of life: Prions and prion-like particles first hypothesis. *Life* **2021**, *11*, 872. [CrossRef] [PubMed]
16. Crane, M.A.; Nair-Desai, S.; Gemmil, A.; Romley, J.A.; Probasco, J.C. Change in epidemiology of Creutzfeldt-Jakob disease in the US, 2007–2020. *JAMA Neurol.* **2024**, *81*, 195–197. [CrossRef] [PubMed]
17. Banerjee, G.; Farmer, S.F.; Hyare, H.; Jaunmuktane, Z.; Mead, S.; Ryan, N.S.; Schott, J.M.; Werring, D.J.; Rudge, P.; Collinge, J. Iatrogenic Alzheimer's disease in recipients of cadaveric pituitary-derived growth hormone. *Nat. Med.* **2024**, *30*, 394–402. [CrossRef] [PubMed]
18. Seo, D.-O.; Holtzman, D.M. Current understanding of the Alzheimer's disease-associated microbiome and therapeutic strategies. *Exp. Mol. Med.* **2024**, *56*, 86–94. [CrossRef] [PubMed]
19. Valles-Colomer, M.; Blanco-Míguez, A.; Manghi, P.; Asnicar, F.; Dubois, L.; Golzato, D.; Armanini, F.; Cumbo, F.; Huang, K.D.; Manara, S.; et al. The person-to-person transmission landscape of the gut and oral microbiomes. *Nature* **2023**, *614*, 125–135. [CrossRef] [PubMed]
20. Deo, P.N.; Deshmukh, R. Oral microbiome: Unveiling the fundamentals. *J. Oral Maxillofac. Pathol.* **2019**, *23*, 122–128. [CrossRef] [PubMed]
21. Fleury, V.; Zekeridou, A.; Lazarevic, V.; Gaïa, N.; Giannopoulou, C.; Genton, L.; Cancela, J.; Girard, M.; Goldstein, R.; Bally, J.F.; et al. Oral dysbiosis and inflammation in Parkinson's disease. *J. Park. Dis.* **2021**, *11*, 619–631. [CrossRef] [PubMed]
22. Dang, A.T.; Marsland, B.J. Microbes, metabolites, and the gut-lung axis. *Mucosal Immunol.* **2019**, *12*, 843–850. [CrossRef] [PubMed]
23. Blum, J.; Silva, M.; Byrne, S.J.; Butler, C.A.; Adams, G.G.; Reynolds, E.C.; Dashper, S.G. Temporal development of the infant oral microbiome. *Crit. Rev. Microbiol.* **2022**, *48*, 730–742. [CrossRef] [PubMed]

24. Samara, P.; Athanasopoulos, M.; Athanasopoulos, I. Unveiling the enigmatic adenoids and tonsils: Exploring immunology, physiology, microbiome dynamics, and the transformative power of surgery. *Microorganisms* **2023**, *11*, 1624. [CrossRef] [PubMed]
25. Martel, J.; Chang, S.H.; Ko, Y.F.; Hwang, T.L.; Young, J.D.; Ojcious, D.M. Gut barrier disruption and chronic disease. *Trends Endocrin. Metab.* **2022**, *33*, 247–265. [CrossRef] [PubMed]
26. Spragge, F.; Bakkeren, E.; Jahn, M.T.; Araujo, E.B.N.; Pearson, C.F.; Wang, X.; Pankhurst, L.; Cunrath, O.; Foster, K.R. Microbiome diversity protects against pathogens by nutrient blocking. *Science* **2023**, *382*, eadj3502. [CrossRef] [PubMed]
27. Turnbaugh, P.J.; Ley, R.E.; Mahowald, M.A.; Magrini, V.; Mardis, E.R.; Gordon, J.I. An obesity-associated gut microbiome with increased capacity for energy harvest. *Nat. Cell Biol.* **2006**, *444*, 1027–1031. [CrossRef] [PubMed]
28. Burberry, A.; Wells, M.F.; Limone, F.; Couto, A.; Smith, K.S.; Keaney, J.; Gillet, G.; van Gastel, N.; Wang, J.-Y.; Pietilainen, O.; et al. C9orf72 suppresses systemic and neural inflammation induced by gut bacteria. *Nature* **2020**, *582*, 89–94. [CrossRef] [PubMed]
29. Larsson, D.G.J.; Flach, C.-F. Antibiotic resistance in the environment. *Nat. Rev. Microbiol.* **2022**, *20*, 257–269. [CrossRef] [PubMed]
30. Reese, A.T.; Chadaideh, K.S.; Diggins, C.E.; Schell, L.D.; Beckel, M.; Callahan, P.; Ryan, R.; Thompson, M.E.; Carmody, R.N. Effects of domestication on the gut microbiota parallel those of human industrialization. *eLife* **2021**, *10*, e60197. [CrossRef] [PubMed]
31. Blaser, M.J.; Falkow, S. What are the consequences of the disappearing human microbiota? *Nat. Rev. Microbiol.* **2009**, *7*, 887–894. [CrossRef] [PubMed]
32. Hung, I.F.N.; Wong, B.C.Y. Assessing the risks and benefits of treating *Helicobacter pylori*. *Ther. Adv. Gastroenterol.* **2009**, *2*, 141–147. [CrossRef] [PubMed]
33. Sehgal, K.; Khanna, S. Gut microbiome and *Clostridioides difficile* infection: A closer look at the microscopic interface. *Therap. Adv. Gastroenterol.* **2021**, *14*, 1756284821994736. [CrossRef] [PubMed]
34. Tariq, R.; Pardi, D.S.; Bartlett, M.G.; Khanna, S. Low cure rates in controlled trials of fecal microbiota transplantation for recurrent *Clostridium difficile* infection: A systematic review and meta-analysis. *Clin. Infect. Dis.* **2019**, *68*, 1351–1358. [CrossRef] [PubMed]
35. Brown, A.C.; Valliere, A. Probiotics and medical nutrition therapy. *Nutr. Clin. Care* **2004**, *7*, 56–68. [PubMed]
36. Pratt, C.; Campbell, M.D. The effect of bifidobacterium on reducing symptomatic pain in patients with irritable bowel syndrome: A systematic review. *Probiotics Antimicrob. Proteins* **2020**, *12*, 834–839. [CrossRef] [PubMed]
37. Khalesi, Z.; Bellisimo, N.; Vandelanotte, C.; Williams, S.; Stanley, D.; Irwin, C. A review of probiotic supplementation in healthy adults: Helpful or hype? *Eur. J. Clin. Nutr.* **2019**, *73*, 24–37. [CrossRef] [PubMed]
38. Kothari, D.; Patel, S.; Kim, S.-K. Probiotic supplements might not be universally-effective and safe.: A review. *Biomed. Pharmacother.* **2019**, *111*, 537–547. [CrossRef] [PubMed]
39. Brüssow, H. Problems with the concept of gut microbiota dysbiosis. *Microb. Biotechnol.* **2019**, *13*, 423–434. [CrossRef] [PubMed]

40. Tao, R.; Zong, G.; Pan, Y.; Li, H.; Cheng, P.; Deng, R.; Chen, W.; Wang, A.; Xia, S.; Tang, W.; et al. *Clostridium butyricum* and *Clostridium tyrobutyricum*: Angel or devil for necrotizing enterocolitis? *mSystems* **2023**, *8*, e0073223. [CrossRef] [PubMed]
41. Segre, J.A. What does it take to satisfy Koch's postulates two centuries later? Microbial genomics and *Propionibacteria acnes*. *J. Investig. Dermatol.* **2013**, *133*, 2141–2142. [CrossRef] [PubMed]
42. Bouchard, C.; Tremblay, A.; Després, J.-P.; Nadeau, A.; Lupien, P.J.; Thériault, G.; Dussault, J.; Moorjani, S.; Pinault, S.; Fournier, G. The response to long-term overfeeding in identical twins. *N. Engl. J. Med.* **1990**, *322*, 1477–1482. [CrossRef] [PubMed]
43. Silventoinen, K.; Sammalisto, S.; Perola, M.; Boomsma, D.I.; Cornes, B.K.; Davis, C.; Dunkel, L.; De Lange, M.; Harris, J.R.; Hjelmborg, J.V.; et al. Heritability of body height: A comparative study of twin cohorts in eight countries. *Twin Res.* **2003**, *6*, 399–408. [CrossRef] [PubMed]
44. Wood, A.; Esko, T.; Yang, J.; Vedantam, S.; Pers, T.H.; Gustafsson, S.; Chu, A.Y.; Estrada, K.; Luan, J.; Kutalik, Z.; et al. Defining the role of common variation in the genomic and biological architecture of adult human height. *Nat. Genet.* **2014**, *46*, 1173–1186. [CrossRef] [PubMed]
45. Neel, J.V. Diabetes mellitus: A "thrifty" genotype rendered detrimental by progress? *Am. J. Hum. Genet.* **1962**, *14*, 353–362. [PubMed]
46. Wang, G.; Speakman, J.R. Analysis of positive selection at single nucleotide polymorphisms associated with body mass index does not support the "thrifty gene" hypothesis. *Cell Metab.* **2016**, *24*, 531–541. [CrossRef] [PubMed]
47. Chaufin, C.; Jay, J. The "missing heritability" of common disorders: Should health researchers care? *Int. J. Health Serv.* **2013**, *43*, 281–303. [CrossRef] [PubMed]
48. Arango, C. Candidate gene associations studies in psychiatry: Time to move forward. *Eur. Arch. Psychiatry Clin. Neurosci.* **2017**, *267*, 1–2. [CrossRef] [PubMed]
49. LeBlanc, J.G.; Chain, F.; Martin, R.; Bermùndez-Humarán, L.G.; Courau, S.; Langella, P. Beneficial effects on host energy metabolism of short-chain fatty acids and vitamins produced by commensal and probiotic bacteria. *Microb. Cell Factories* **2017**, *16*, 79. [CrossRef] [PubMed]
50. Torres-Fuentes, C.; Schellenkens, H.; Dinan, T.G.; Cryan, J.F. The microbiota-gut-brain axis in obesity. *Lancet Gastroenterol. Hepatol.* **2017**, *2*, 747–756. [CrossRef] [PubMed]
51. Calvo-Barreiro, L.; Zhang, L.; Abdel-Rahman, S.A.; Naik, S.P.; Gabr, M. Gut microbial-derived metabolites as immune modulators of T helper 17 and regulatory T cells. *Int. J. Mol. Sci.* **2023**, *24*, 1806. [CrossRef] [PubMed]
52. Dinan, T.G.; Cryan, J.F. Gut microbiota: A missing link in psychiatry. *World Psychiatry* **2020**, *19*, 111–112. [CrossRef] [PubMed]
53. Reigstad, C.S.; Salmonson, C.E.; Rainey, J.F., III; Szurszewski, J.H.; Linden, D.R.; Sonnenburg, J.L.; Farrugia, G.; Kashyap, P.C. Gut microbes promote colonic serotonin production through an effect of short-chain fatty acids on enterochromaffin cells. *FASEB J.* **2005**, *29*, 1395–1403. [CrossRef] [PubMed]
54. Xiong, R.G.; Zhou, D.D.; Wu, S.X.; Huang, S.Y.; Saimaiti, A.; Yang, Z.J.; Shange, A.; Zhao, C.N.; Gan, R.Y.; Li, H.B. Health benefits and side effects of short-chain fatty acids. *Foods* **2022**, *11*, 2863. [CrossRef] [PubMed]

55. Fusco, W.; Lorenzo, M.B.; Cintoni, M.; Porcari, S.; Rinninela, E.; Kaitsas, F.; Lener, E.; Mele, M.C.; Gasbarrini, A.; Collado, M.C.; et al. Short-chain fatty-acid-producing bacteria: Key components of the human gut microbiota. *Nutrients* **2023**, *15*, 2211. [CrossRef] [PubMed]
56. Sonnenburg, E.D.; Smits, S.A.; Tikhonov, M.; Higginbottom, S.A.; Wingreen, N.S.; Sonnenburg, J.L. Diet-induced extinctions in the gut microbiota compound over generations. *Nature* **2016**, *529*, 212–215. [CrossRef] [PubMed]
57. O'Keefe, S.J. The association between dietary fibre deficiency and high-income lifestyle-associated diseases: Burkitt's hypothesis revisited. *Lancet Gastroenterol. Hepatol.* **2019**, *4*, 984–996. [CrossRef] [PubMed]
58. Burkitt, D.P. Some diseases characteristic of modern western civilization. *Br. Med. J.* **1973**, *1*, 274–278. [CrossRef] [PubMed]
59. Schnorr, S.L.; Candela, M.; Rampelli, S.; Centanni, M.; Consolandi, C.; Basaglia, G.; Turroni, S.; Biagi, E.; Peano, C.; Severgnini, M.; et al. Gut microbiome of the Hadza hunter–gatherers. *Nat. Commun.* **2014**, *5*, 3654. [CrossRef] [PubMed]
60. Carter, M.M.; Olm, M.R.; Merrill, B.D.; Dahan, D.; Tripathi, S.; Spencer, S.P.; Yu, F.B.; Jain, S.; Neff, N.; Jha, A.R.; et al. Ultra-deep sequencing of Hadza hunter–gatherers recovers vanishing gut microbes. *Cell* **2023**, *186*, 3111–3124. [CrossRef] [PubMed]
61. Hidalgo, G.; Marini, E.; Sanchez, W.; Contreras, M.; Estrada, I.; Comandini, O.; Buffa, R.; Magris, M.; Dominguez-Bello, M.G. The nutrition transition in the Venezuelan Amazonia: Increased overweight and obesity with transculturation. *Am. J. Hum. Biol.* **2014**, *26*, 710–712. [CrossRef] [PubMed]
62. Clemente, J.C.; Pehrsson, E.C.; Blaser, M.J.; Sandhu, K.; Gao, Z.; Wang, B.; Magris, M.; Hidalgo, G.; Contreras, M.; Noya-Alarsón, Ó.; et al. The microbiome of uncontacted Amerindians. *Sci. Adv.* **2015**, *1*, e1500183. [CrossRef] [PubMed]
63. Schaan, A.P.; Sarquis, D.; Cavalcante, G.C.; Magalhães, L.; Sacuena, E.R.P.; Costa, J.; Fonseca, D.; Mello, V.J.; Guerreiro, J.F.; Ribeiro-dos-Santos, A. The structure of Brazilian Amazonian gut microbiomes in the process of urbanisation. *npj Biofilms Microbiomes* **2021**, *7*, 65. [CrossRef] [PubMed]
64. Kaplan, H.; Thompson, R.C.; Trumble, B.C.; Wann, L.S.; Allam, A.H.; Beheim, B.; Frohlich, B.; Sutherland, M.L.; Sutherland, J.D.; Stieglitz, J.; et al. Coronary atherosclerosis in indigenous South American Tsimane: A cross sectional cohort study. *Lancet* **2017**, *389*, 1730–1739. [CrossRef] [PubMed]
65. Irimia, A.; Chaudhari, N.N.; Robles, D.J.; Rostowsky, K.A.; Maher, A.S.; Chowdhury, N.F.; Calvillo, N.F.; Ngo, V.; Gatz, M.; Mack, W.J.; et al. The indigenous South American Tsimane exhibit relatively modest decrease in brain volume with age despite high systemic inflammation. *J. Gerontol. A Biol. Sci. Med. Sci.* **2021**, *76*, 2147–2155. [CrossRef] [PubMed]
66. Cao, T.; Cortez, E.C.; Miyamoto, M.I.; Cummings, D.; Trumble, B.C.; Stieglitz, J.; Thompson, R.C.; Kaplan, H.; Thomas, G.S.; Duprez, D.A.; et al. Abstract 13719: Minimal and delayed age-related increase of arterial stiffness among Tsimane forager-horticulturalists. *Circulation* **2023**, *148*, A13719. [CrossRef]

67. Sprockett, D.D.; Martin, M.; Costello, E.L.; Burns, A.R.; Holmes, S.P.; Gurven, M.D.; Relman, D.A. Microbiota assembly, structure, and dynamics among Tsimane horticulturalists of the Bolivian Amazon. *Nat. Commun.* **2020**, *11*, 3772. [CrossRef] [PubMed]
68. Dixson, A.F.; Dixson, B.J. Venus figurines of the early paleolithic: Symbols of fertility or attractiveness? *J. Anthropol.* **2011**, *2011*, 569120. [CrossRef]
69. Wibowo, M.C.; Yang, Z.; Borry, M.; Hübner, A.; Huang, K.D.; Tierney, B.T.; Zimmerman, S.; Barajas-Olmos, F.; Contreras-Cubas, C.; García-Ortiz, H.; et al. Reconstruction of ancient microbial genomes from the human gut. *Nature* **2021**, *594*, 234–239. [CrossRef] [PubMed]
70. Elizabeth, L.; Machado, P.; Zinöcker, M.; Baker, P.; Lawrence, M. Ultra-processed foods and health outcomes: A narrative review. *Nutrients* **2020**, *12*, 1955. [CrossRef] [PubMed]
71. Gearhardt, A.N.; DiFeliceantonio, A.G. Highly processed foods can be considered addictive substances based on established scientific criteria. *Addiction* **2023**, *118*, 581–779. [CrossRef] [PubMed]
72. Samuthpongtorn, C.; Nguyen, L.H.; Okereke, O.I.; Wang, D.D.; Song, M.; Chan, A.T.; Mehta, R.S. Consumption of ultraprocessed food and risk of depression. *JAMA Netw. Open* **2023**, *6*, e2334770. [CrossRef] [PubMed]
73. Hall, K.D.; Heymsfield, S.B.; Kemnitz, J.W.; Klein, S.; Schoeller, D.A.; Speakman, J.R. Energy balance and its components: Implications for body weight regulation. *Am. J. Clin. Nutr.* **2012**, *95*, 989–994. [CrossRef] [PubMed]
74. Ludwig, D.S.; Aronne, L.J.; Astrup, A.; de Cabo, R.; Cantley, L.C.; Friedman, M.I.; Heymsfield, S.B.; Johnson, J.D.; King, J.C.; Krauss, R.M.; et al. The carbohydrate-insulin model: A physiological perspective on the obesity pandemic. *Am. J. Clin. Nutr.* **2021**, *114*, 1873–1885. [CrossRef] [PubMed]
75. Hall, K.D. A review of the carbohydrate-insulin model of obesity. *Eur. J. Clin. Nutr.* **2017**, *71*, 323–326. [CrossRef] [PubMed]
76. Ludwig, D.S.; Sørensen, T.I.A. An integrated model of obesity pathogenesis that revisits causal direction. *Nat. Rev. Endocrinol.* **2022**, *18*, 261–262. [CrossRef] [PubMed]
77. Sievenpiper, J.L. Low-carbohydrate diets and cardiometabolic health: The importance of carbohydrate quality over quantity. *Nutr. Rev.* **2020**, *78* (Suppl. S1), 69–77. [CrossRef] [PubMed]
78. Simopoulos, A.P. *The Healthiest Diet for You: Scientific Aspects*; MDPI: Basel, Switzerland, 2022.
79. Innes, J.K.; Calder, P.C. Omega-6 fatty acids and inflammation. *Prostaglandins Leukot. Essent. Fat. Acids* **2018**, *132*, 41–48. [CrossRef] [PubMed]
80. Hébert, J.R.; Shivappa, N.; Wirth, M.D.; Hussey, J.R.; Hurley, T.G. Perspective: The dietary inflammatory index (DII)–lessons learned, improvement made, and future directions. *Adv. Nutr.* **2019**, *10*, 185–195. [CrossRef] [PubMed]
81. Mozaffarian, D. Perspective: Obesity-an unexplained epidemic. *Am. J. Clin. Nutr.* **2022**, *115*, 1445–1450. [CrossRef] [PubMed]
82. Rose, C.; Parker, A.; Jefferson, B.; Cartmell, E. The characterisation of feces and urine: A review of the literature to inform advanced treatment technology. *Crit. Rev. Environ. Sci. Technol.* **2015**, *45*, 1827–1879. [CrossRef] [PubMed]

83. Forootan, M.; Bagheri, N.; Darvishi, M. Chronic constipation. *Medicine* **2018**, *97*, e10631. [CrossRef] [PubMed]
84. Vandeputte, D.; Falony, G.; Veira-Silva, S.; Tito, R.; Joossens, M.; Raes, J. Stool consistency is strongly associated with gut microbiota richness and composition, enterotypes and bacterial growth rates. *Gut* **2016**, *65*, 57–62. [CrossRef] [PubMed]
85. Keesey, R.E.; Hirvonen, M.D. Body weight set-points: Determination and adjustment. *J. Nutr.* **1997**, *127*, 1875S–1883S. [CrossRef] [PubMed]
86. Careau, V.; Halsey, L.G.; Pontzer, H.; Ainslie, P.N.; Andersen, L.F.; Anderson, L.J.; Arab, L.; Baddou, I.; Bedu-Addo, K.; Blaak, E.E.; et al. Energy compensation and adiposity in humans. *Curr. Biol.* **2021**, *31*, 1–8. [CrossRef] [PubMed]
87. Protsiv, M.; Ley, C.; Lankester, J.; Hastie, T.; Parsonnet, J. Decreasing human body temperature in the United States since the Industrial Revolution. *eLife* **2020**, *9*, e49555. [CrossRef] [PubMed]
88. Levine, J.A. Non-exercise activity thermogenesis (NEAT). *Nutr Rev.* **2004**, *62*, S82–S97. [CrossRef] [PubMed]

Part 2. The Evolutionary Background of the Microbiome

Science, Old and New: Just as you cannot make an omelette without breaking eggs, so old ideas have to give way to new and, for a while at least, "new science" can seem very close to anti-science. Indeed, the only reliable way to distinguish the two is with the aid of hindsight: anti-science will have been believed regardless of the evidence, while new ideas will have been tested by experiments. Understanding the role of microeukaryotes, not just bacteria, will hopefully allow a greater range of options, not only for understanding disease, but also for its amelioration, and even prevention in the future. An example from chemistry may illustrate the point: just as it is not possible to study the active conformations of peptides and proteins without reference to the water molecules in which they operate, so it is not possible to fully understand multicellular organisms without reference to the microbe-rich world in which they operate. Although, sadly, some things must remain hypothetical. It may be that the development of the vertebrate body/microbiome "holobiont" can be summarised in just four basic steps from an original abiotic environment:

1. *Emergence*: Although the exact mechanism remains unclear, the replicable chemistry that is the prokaryotes, the Bacteria and the Archaea, developed (emerged) out of an abiotic environment.
2. *Endosymbiosis*: At some point, one prokaryote began to live inside the cell of another, eventually adjusting their respective genes to become an organelle inside what has become a unicellular eukaryote. Note that this microeukaryote would, somehow, have contributed its enhanced skills to the microbial community in which it formed.
3. *Inversion*: As eukaryotic cells gained the ability to cooperate with one another, so a portion of the original microbial community may have remained inside the newly formed multicellular creature, with both sides adjusting their genes in a higher-level version of endosymbiosis. A form of "domesticated" unicellular eukaryote may have developed to defend the newly formed animal during this stage, eventually becoming immune system cells.
4. *Elongation*: Under the influence of Darwinian selection, some of the resulting creatures developed into the vertebrates, the primary microbial community remaining within the lower part of their intestine so as to allow its transfer to the next generation. It is the combination of body and microbial community acting as a single entity that is the holobiont, while some cells may have retained their (technically) external status as microbial sentinel cells.

5. The Holobiont

5.1. The Emergence of Life

The fact that life somehow "emerged" from an abiotic mixture of minerals and molecules remains one of the greatest mysteries of the natural world [1], as relevant to the icy moons found in the solar system as to the Hadean Earth [2]. It is possible—even likely—that mobile genetic elements and the nearly live viruses [3], may have preceded the first cells. Indeed, it has even been suggested that viruses may have populated space itself before finally gaining a foothold on Earth [4]. What eventually "emerged" from the abiotic environment of the early Earth were the prokaryotes, the first recognisable forms of single-celled life, the Bacteria and the Archaea. Although there are many traits that distinguish these two in the modern world, one of the most significant is the type of chemical connection in their lipid structures: "ester" and "ether," respectively [5]. In purely chemical terms, ester bonds are relatively easy to form but hydrolytically unstable, while ether linkages are the opposite; harder to form, maybe, but more stable under extreme conditions. Sadly, however, the superficially attractive idea that the Archaea formed first in the Hadean aeon, the hot, hell-like early period of Earth's history, cannot be confirmed.

5.2. Darwinism

It is interesting to note the influence of society on science. From the middle of the 19th century, the dominant idea was the survival of the fittest and, likewise, by the mid-20th century, the genetics revolution had hardly altered that concept, the gene replacing the species as the unit of survival [6]. The cataclysmic struggles of the Second World War, and the subsequent Cold War, simply reinforced these concepts, as all the contenders naturally thought that they were in the right. Overall, this background story fitted nicely with the idea of "good" humans versus "bad" bacteria and, indeed, it also fitted with US President Nixon's 1971 declaration of "war" on cancer. Of course, fifty years after Nixon's declaration, it is clear that the war on cancer is being lost, with a new epidemic becoming apparent in the young [7]. Darwinism is not enough. A new theory is urgently needed.

5.3. Carl Woese

Carl Richard Woese was born in 1928 and, among many contributions to biology, he helped to establish the two "domains" of the prokaryotes: Archaea and Bacteria, which can be characterised using the class of molecular structures referred to as "16S rRNA" (specifically, the RNA component of the 30S subunit of a prokaryotic ribosome) [8]. Later, these two were combined with the Eukarya (sometimes spelt Eucarya) in a three-domain system which includes our own eukaryotic body cells [9]. Later still, Woese acknowledged the fact that Darwinian-type survival of the fittest applies primarily to the multicellular eukaryotes, which pass on their genes "vertically" from one generation to the next. By contrast, the single-celled prokaryotes may pass on their characteristics horizontally, i.e., encoded by mobile genetic elements [10]. Woese coined the phrase "Darwinian threshold" to illustrate the significance of these findings, that the stability of vertical gene transfer, above the threshold, can be complemented by the flexibility of horizontal gene transfer, which operates below the threshold.

5.4. Mobile Genetic Elements

The concept of mobile genetic elements as pathogen modifiers in microbial populations is of interest [11]. More recent concerns about the development of resistance to antibiotics has resulted in increased focus on the role of the "bacteriophage" viruses in horizontal gene transfer [12]; however, it is important to note that other means of deactivating bacteria may also lead to functional uptake in their neighbours, as observed by Griffith's studies on pneumonia in the early part of the 20th century [13]. A process of bacteriophage-accelerated nutrient cycling has been described as occurring within defined boundaries. Although originally reported within layers of the sea [14] it is likely that the same principle could be applied to the intestinal microbiome in that, along with the macro- and micro-nutrients of food, microbes and their related antigens and "phage" viruses will also be found, as illustrated by Figure 5.

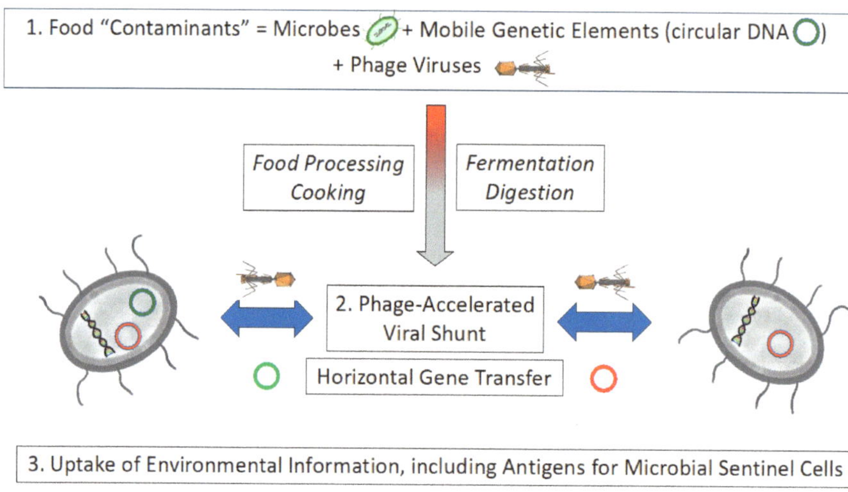

Figure 5. Environmental information. Box 1. In addition to the micro- and macro-nutrients of food there are a range of "contaminants" in the form of microbes and their attendant mobile genetic elements and phage viruses. Food processing and digestion will change these components, and the immune system will engage with their antigens, passing on relevant information to the brain and gut. Box 2. When in the large intestine, mobile genetic elements will be uploaded into the diverse microbiome, accelerated by phage action according to the viral shunt mechanism. Box 3. At the appropriate time, the full repertoire of immune system information will be collected in the pregnant female, ready to be passed on to the next generation. Source: Adapted from Smith et al. [15], used with permission.

5.5. *Lynn Margulis*

Lynn Petra Alexander was born in 1938 and, initially as Lynn Sagan, later Margulis, she began to compare bacteria and archaea with the eukaryotes. Though she concluded that the Eukarya were formed by the symbiotic fusion of representatives of these two prokaryotes, many battles had to be fought before this work was finally published, in 1967, as "On the Origin of Mitosing Cells" [16]. (The word "mitosis" represents the process of cellular division within the Eukarya.) Because it went against the prevailing neo-Darwinian orthodoxy of the day, this work was ignored for many years, only becoming recognised as a breakthrough (and after more "free and frank" academic battles) as the genetic evidence accumulated in its favour. The basic process is an internal growing together, an endosymbiosis, with two free-living microbes becoming entangled, altering their genomes so as to complement one another. In this way, the "invader" became an organelle, a unit within the cell, while the enhanced cell itself became what is now known as a eukaryote, a Greek-derived word stemming from its resemblance to a nut surrounding its kernel (the "prokaryotic" bacteria and archaea are so named because they do not possess these kernel-like organelles). The fundamental role of this "kernel" is as the energy-producing mitochondrion, a flexible subunit, capable of moving from one cell to another by the normal processes of horizontal gene transfer, with consequences that are still being uncovered [17].

5.6. *Symbiosis and Skillsets*

On the grounds that you can never have enough of a good thing, these endosymbiotic processes have been repeated on a number of occasions, one example being the ancient uptake of photosynthetic cyanobacteria into eukaryotic cells, rearranging their genetic machinery to become chloroplasts, the major source of chemical energy at the lowest trophic level: the bottom of the food chain [18]. Similarly, it has recently been shown that a nitrogen-fixing bacterium is busily incorporating itself inside the splendidly named alga *Braarudosphaera bigelowii*, becoming what has been described as a "nitroplast" [19]. Interestingly, the process of organelle formation has been detected in a part of the Norwegian Sea called Loki's Castle, even though the experimental work required to prove this is remarkably challenging [20]. However it took place, these unicellular eukaryotes (also called microeukaryotes), whether free-living or parasitic, have significantly enhanced characteristics compared to the prokaryotes. As an example, the parasite *Toxoplasma gondii* lives inside the cells of the nervous system of rodents, interfering with their function so as to prevent escape when threatened by felines, the host for the next stage in their lifecycle [21]. It is important to note, however, that endosymbiosis is not necessarily mutually beneficial [22], and that theoretical work on what has been termed "host-beneficial intracellular infections" will never be completely finished [23].

5.7. *Multicellularity*

Discounting colonies such as biofilms, one of the many things that eukaryotes do better than prokaryotes lies in their ability to form multicellular organisms [24].

Although early attention was drawn to the Cambrian period of the Palaeozoic era, the time when recognisable animals first developed, a group of 1950s schoolchildren were surprised to discover what looked like a fossil from Precambrian rocks near Leicester in the United Kingdom [25]. Interestingly, one of the imprints in the rocks from that period, a creature named *Dickinsonia*, was found to be associated with steroid residues, confirming it as a variety of animal from about 600 million years ago [26]. It is important to note that the transition to multicellularity is a two-way process [27], of which the best-studied example is the yeasts [28], potentially controlled by a collective process called quorum sensing [29]. Another example is the unattractively named "slime moulds" exhibiting a switch between single- and multi-celled behaviour, but also offering valuable insights into collective decision-making [30].

5.8. Vertebrates

A major advance was taken with the development of an elongated body plan, the first chordates being found in the Cambrian period [31]. In turn, these were followed by the vertebrate jawless fishes, including a species resembling the modern hagfish, found in deposits associated with the ancient Tethys Sea [32]. Modern hagfish are well-studied specialised flesh-burrowing scavengers with no sign of a specific microbe repository similar to the large intestine of mammals [33], although they do utilise a chitin-based peritrophic membrane, apparently in order to exclude excessive microbial interaction [34], and in this they are reminiscent of the invertebrates. However, the jawless hagfish and lampreys do possess a type of adaptive immunity, theoretically allowing them to co-exist with bacteria [35] although there are no confirmatory studies relating specifically to the hagfish [33]. Further along the vertebrate evolutionary pathway, the development of the amniotic egg allowed vertebrate animals an aqueous-independent existence on land [36], endothermy having developed to cope with the more extreme hot–cold cycles out of the water [37]. This ancient "elongation" seems to have spatially separated the two functions of the holobiont: the microbial component of the immune system on the one hand, and the animal brain and nervous system on the other, and it could be that this separation has left its mark in the involvement of the gut–brain axis in today's epidemics of non-communicable disease. Specifically, it seems that there is a "gut feeling" involved at an emotional level in the human decision-making process, perhaps because a pre-vertebrate "assessment of significance" originally included a role for the microbial component. In fact, developmental difficulties may also extend to the mating inability of a high proportion of captive male pandas, who show significant variation in *Clostridium* species between those animals either capable or apparently incapable of natural mating [38]. In a sense, captive animals have been domesticated, a process that has been reported to have a microbiome-degrading effect somehow resembling industrialisation [39]. Indeed, Denis Burkitt's observations on the absence of disease in traditional societies perhaps imply that humans in "Modern Western Civilisation" have themselves been "domesticated" by industrialisation [40].

5.9. Holobiont

Fresh from her success in understanding the symbiotic genesis of the eukaryotes, now-Professor Lynn Margulis looked at symbiosis on the grander scale, using the term "holobiont" to indicate a number of widely different organisms acting together as a driver of speciation [41]. While a great deal of work has focused on the interaction of the economically important insects with bacteria [42], it could be that they are too readily occupied by microbial entities, somewhat complicating the picture [43]. By contrast, it seems that our bacteria have diversified along with humans across Asia, Africa, and Europe, perhaps indicating a greater stability when associated with the vertebrate intestinal microbiome [44]. Sadly, most studies refer to bacteria, but archaea are also found in the human microbiome, an entire domain of life with no known pathogenic representatives [45], and a suggestion has been made that microbial eukaryotes are vital to the microbiome, constituting a "missing link" between prokaryotes and animals [46], possibly working together through the immune system [47]. One potential route to the vertebrate holobiont is sketched out in Figure 6, albeit covering billions of years, in which the prokaryote microbial community that gave rise to the first eukaryote cell was eventually reduced to its essentials, surrounded by a multicellular skin/armour, and equipped with the apparatus of survival: brain and nervous system; sensory organs, mobility, and teeth. Of course, there is an analogy with the formation of the eukaryote itself, as the bacterium, surrounded by its host cell, lost some aspects of its genetic abilities in favour of the whole. Accordingly, in one sense, the microbiome can be looked on as organ of the body, perhaps most closely associated with the immune system, although adults seem to remain superficially healthy after colectomy and, therefore, without a normal intestinal microbiome [48]. In another sense, however, as its role extends across the generations, perhaps the fully functioning microbiome can be looked on as an organ of the species, rather than of the individual.

5.10. Semi-Independent Cells

In spite of decades of work, it seems that our knowledge of the immune system is hardly out of its infancy. One favourite line of approach is to investigate model systems, such as the nematode, *Caenorhabditis elegans*, as a simple stand-in for the innate immune system [49]. Study of its bacteria, fungi and viruses has been initiated, but primarily considered to be, at best, near-pathogens [50]. The significance of their so-called commensals is also of interest, noting their specific strain-relevance to the prevention of bacterial infection [51]. More recent work acknowledges that such commensals may actually be part of the nematode microbiome, and associates them with fitness and longer lifespan [52], much as in humans. The truth remains to be seen, but it is notable that these studies stick to bacteria and do not look for unicellular eukaryotes. In humans, dendritic cells are known to communicate between the innate and adaptive systems as so-called "sentinel cells" (antigen-presenting cells), contribute to immunological tolerance [53] and, it is interesting to note, may prowl around bodily tissues as if they were independent unicellular eukaryotes [54]. For a number of years now, it has been recognised that protists, including *Blastocystis* strains, not only occupy many different vertebrate hosts [55], but are also "prevalent and diverse" members of the human microbiome, potentially without causing any disease [56]. Likewise, a more recent review accepts that at least some subtypes of *Blastocystis* are well distributed [57], may indeed be part of the healthy microbiome [58], and that some seem to gain benefit from dietary plant spices, as well as being adversely affected by antibiotics [59].

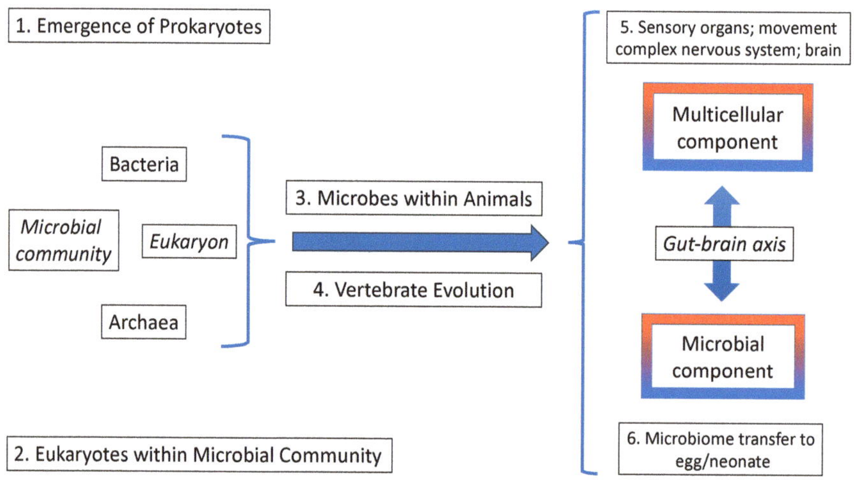

Figure 6. A summary of vertebrate evolution. Left-hand side: This represents the supposed initial stages of evolution as (as stated in Box 1) the prokaryote Bacteria and Archaea somehow emerged from an abiotic environment. In turn, (Box 2) the first of the Eukarya arose from endosymbiosis of representatives of the prokaryotes. Central arrow: It is possible to imagine that an initial multicellular entity enclosed a residual microbial community to afford (Box 3) an animal holobiont. Eventually, (Box 4), subsequent Darwinian evolution favoured an elongated shape, giving rise to the early vertebrates. Right-hand side: Further evolution (Box 5) allowed the development of the multicellular body as a food-seeking machine while (Box 6) the primary microbial community, the microbiome, remained in the lower digestive tract in order to engage with the next generation via apparently accidental contamination. Source: Figure by the Author.

5.11. Vertebrate Microbiome Inheritance

The extension of the vertebrate body plan in the terrestrial animals potentially leads to difficulties in the "dual inheritance": the genetic inheritance of the animal on the one hand, and the environmental information-carrying microbiome, including any microbial sentinel cells, on the other. In mammals, the birth canal forms alongside the anus, maximising the chance of microbial transfer, albeit seemingly by accident, whereas cloacal contact occurs extensively during the laying season of birds, allowing extensive exchange of microbes [60]. In fact, it has been estimated that the brood-parasitic cowbird *Molothrus ater*, along with their passerine hosts, exhibit a greater microbial diversity than all the mammalian species put together [61]. While the provision of milk virtually defines mammals, it is interesting to note that birds of the economically important pigeon family also produce "milk," but from the crop of both male and female. Significantly, both pigeon crop milk [62] and human breast milk [63] contain antibodies against potential pathogens, but they also contain bacteria, particularly *Lactobacilli* and *Bifidobacteria* [63]. Although the exact function of these bacteria remains unknown, one possibility is that they are harmless commensals, tolerated specifically for the purpose of stimulating the immune system of neonate and squab (pigeon chick). If so, the principle would be exactly the same as a cat bringing home a live mouse for her kittens to practise on, i.e., honing their hunting skills.

5.12. *Soft Inheritance*

Ever since Conrad Waddington noticed discrepancies in the standard neo-Darwinism of his time [64], evidence has been accumulating that gene-modifying functions can be inherited across several generations without any change in DNA sequences. This so-called soft inheritance occurs when some environmental signal more or less temporarily stops the genes from working, at least partly, and this has been termed "epigenetic" chemical modification of the machinery of transcription [65]. For example, this question of soft inheritance was raised in an animal study when it was observed that the sweetener aspartame induced anxiety, not only in the murine "users" themselves, but also in subsequent generations [66]. Not everybody was convinced about the direct inheritance of epigenetic markers, however [67], and it has been suggested that microbes transferred between generations account for the transfer of obesity from mother to child [68]. While some researchers have explored the interaction between bacteria and the so-called "epigenome" of animals [69], sadly, little is known for certain. Interestingly, the case has been made for paternal transfer of information, for which the maternal microbiome cannot easily be invoked. It seems feasible that epigenetic changes in the spermatozoa also contribute to the later development of the offspring [70]. The inheritance debate is far from over!

5.13. *Fungi and the Mycobiome*

Like bacteria, fungi are relatively easy to distinguish from the other components of the microbiome and, accordingly, much work has been carried out on the so-called mycobiome, finding that they occupy all of the available niches around the human body [71]. Although smaller in absolute number, fungal diversity can be almost as great as for the bacteria [72], and they have been associated with the development of type 1 diabetes [73]. This last point is important, as studies performed in the presence of microbiome-related disease may not be a true representation of the fully functioning microbiome, implying that the presence of fungi could, indeed, be associated with disease. Significantly, perhaps, it is worth noting that microbial sentinel cells could stem from the same evolutionary precursor as equivalent cells of the innate immune system, clearly a niche that cannot easily be occupied by fungi. In this context, a study has been carried out on the transfer of protozoa and fungi from mother to calf in grazing ruminants (yaks). It seems that protozoa were preferred [74], but sadly, no data were presented on the function of these more readily transferred unicellular eukaryotes.

5.14. *Prehistoric Oral Microbiomes*

One of the best sources of information about the past is the calculus that builds up around the teeth, and it is not surprising that researchers have been able to track the change in diet between Paleolithic hunter–gatherers and late Neolithic farmers from the same region within the Italian peninsula [75]. A similar picture has been derived from Great Britain, in a study of teeth from 253 individuals, covering the period from about 2200 BCE to 1853 CE. As is consistently observed, the trapped microbiota were more diverse than those found in modern samples [76]. The teeth may also preserve clues to what may have caused the loss of so many microbes, as lead and other toxic heavy metals leave characteristic patterns due their effect on dental health [77] and may, therefore, cause damage to the microbiome.

5.15. *Pollution*

While the majority of recent efforts have gone into unravelling the genome, its interaction with the external environment remains unclear, a fact that is brought sharply into focus when considering industrial pollution. The concept of the "exposome" has been developed to assess the relationship between genome and pollution [78], further developed more recently [79], and added to by a list of the "hallmarks of environmental insults", including the changes in microbiome communities [80]. Macroscopic plastic pollution is very visible and has a high profile, with microplastics contaminating our food [81], while plasticisers and heavily halogenated substances have been implicated as endocrine disruptors [82]. However, of all the potential pollutants, while heavy metals may be sequestered, they cannot be degraded, and are therefore more likely to cause irreversible damage to the microbiome. Indeed, it is known that the accumulation of toxic metals in Greenland ice matches the progress of European industrialisation over many years, both with lead [83] and with other heavy metals [84].

5.16. *Heavy Metals*

The bulk of modern-day efforts to understand the aetiology of non-communicable disease focus on our changing foods, either in the mouse model [85], or with respect to the move away from hunter–gatherer lifestyles [86]. Additionally, studies on the bacterial microbiome indicate that heavy metal pollution during the course of industrialisation could be to blame [87]. Toxic metal ores have been mined since Neolithic times [88], and lead residues consistent with cosmetic use have found in ceramic bottles as far back as the 5th millennium BCE [89], well within the time period of the people who made the Venus figurines [90]. Pots containing galena (lead(II) sulphide) have been found in Ancient Egyptian tombs, as have atherosclerotic plaques in the arteries of their mummies [91]. Likewise, the Roman Empire made extensive use of lead but, when the skeleton of a girl was uncovered from a high-status burial in the first-century CE Italian peninsula bearing the unmistakeable signs of coeliac disease, researchers focused on the genetic explanation of "predisposing haplotypes" [92], not realising that there may be a connection with similar diseases in today's extensively polluted world. Much later, the mother of John Bostock could have been using brightly coloured salts of lead and similar metals to beautify herself [93], with consequent problems for the microbiome of her offspring and his subsequent development of hay fever [94]. Interestingly, genes coding for tolerance to heavy-metal poisoning have been found in the microbiomes of people living in heavily contaminated areas of the Brazilian Amazon and, separately, in the Peruvian Andes [95]. Bearing in mind the hypothesised relationship with microbial sentinel cells and the immune system, it may be that microbiome poisoning is more likely if the heavy metals are presented as insoluble microbe-sized particles, rather than an otherwise simple solution.

5.17. Poisoning Animals

Finally, while Burkitt, based on his African experience, states that his Diseases Characteristic of Modern Western Civilization are "rare or unknown in undomesticated animals" [43], the same may not be true in otherwise wild animals living in close proximity to such civilisations. Researchers have noted that English sub-populations of the European hedgehog, *Erinaceus europaeus*, have been in steady decline for many years, most dramatically since the year 2000 [96]. These areas are among the most highly developed, highly polluted parts of the world, with few wild places available for populations under environmental stress to recover. It could be that these hedgehogs are uniquely vulnerable to non-communicable diseases brought about by pollution specifically affecting the microbiome, possibly alongside the poisoning of the body itself. Likewise, insect populations are in decline [97] and, as may be expected, they are also affected by metal toxicity [98].

5.18. Susceptibility to Poisoning

In evolutionary terms, defences against heavy-metal poisoning would have evolved as already established populations of relatively immobile entities became exposed to the weathering of nearby ores. As illustrated in Figure 7, however, it will be plants that will have evolved the most robust defences, presumably based on sequestration, followed by the less mobile browsing invertebrates. Conversely, wide-ranging vertebrates, including hominids, would presumably have had little evolutionary pressure to cope with what was, essentially, a local problem, at least before the current episodes of industrialisation. As the increasing danger posed by pollutants, including heavy metals, is realised, intense efforts to clean up our environment are underway [99]. Although certain classes of microbe can be used for bioremediation [100], it is hard to see how they could survive long enough to do this same job in the microbiome. Needless to say, this is a very new area indeed [101].

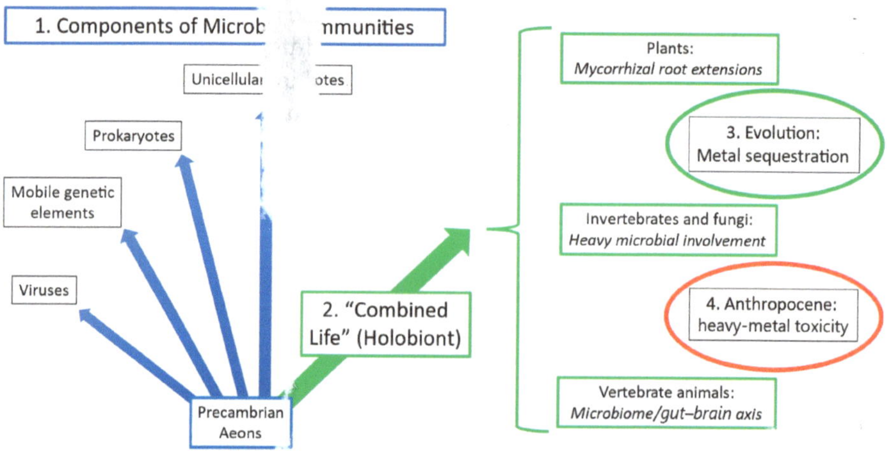

Figure 7. Vulnerability to poisoning. Box 1: This "palm print" diagram represents the basic gene-based categories that have arisen out of the Precambrian aeons, in turn forming the constituents of microbial communities and, indeed, microbiomes. Box 2: This illustrates the basic entities currently found around our planet. Right-hand side, Box 3: The plants and their associated invertebrate herbivores will have been exposed to newly weathered heavy metal ores, allowing them to evolve methods of sequestration. By contrast, the more mobile vertebrates will not have been able to develop such defences and, accordingly, will be vulnerable to the rapid change in toxicity profile observed within the modern, heavily polluted Anthropocene. Source: Figure by the Author.

5.19. Summary

Professors Margulis and Woese helped launch revolutions in the way that we think about evolution, whose consequences are currently being worked through. While the version of symbiosis presented in this chapter is arguable, it does provide a basis upon which to understand non-communicable disease: that the fundamental role of the microbiome is as an accessory to the vertebrate immune system.

References

1. Jheeta, S. Conference Report—Molecules to Microbes. *Science* **2019**, *1*, 1–14.
2. Camprubí, E.; de Leeuw, J.W.; House, C.H.; Raulin, F.; Russell, M.J.; Spang, A.; Tirumalai, M.R.; Westall, F. The emergence of life. *Space Sci. Rev.* **2019**, *215*, 56. [CrossRef]
3. Moelling, K.; Broecker, F. Viruses and evolution-viruses first? A personal perspective. *Front. Microbiol.* **2019**, *10*, 523. [CrossRef] [PubMed]
4. Slijepcevic, P.; Wickramasinghe, N.C. A cosmic virosphere. *Adv. Genet.* **2020**, *106*, 101–107. [PubMed]
5. van Wolferen, M.; Pulschen, A.; Baum, B.; Gribaldo, S.; Albers, S.-V. The cell biology of Archaea. *Nat. Microbiol.* **2022**, *7*, 1744–1755. [CrossRef] [PubMed]

6. Ågren, J.A.; Clark, A.G. Selfish genetic elements. *Genet.* **2018**, *14*, e1007700. [CrossRef] [PubMed]
7. Ugai, T.; Sasamoto, N.; Lee, H.Y.; Ando, M.; Song Tamimi, R.M.; Kawachi, I.; Campbell, P.T.; Giovannucci, E.L.; Weiderpass, E.; et al rly-onset cancer an emerging global epidemic? Current evidence and future implica s. *Nat. Rev. Clin. Oncol.* **2022**, *19*, 656–673. [CrossRef] [PubMed]
8. Woese, C.R.; Fox, G.E. Phylogenetic structure of the prokaryotic domain: The primary kingdoms. *Proc. Natl. Acad. Sci. USA* **1977**, *74*, 5088–5090. [CrossRef] [PubMed]
9. Woese, C.R.; Kandler, O.; Wheelis, M.L. Towards a natural system of organisms: Proposal for the domains Archaea, Bacteria, and Eucarya. *Proc. Natl. Acad. Sci. USA* **1990**, *87*, 4576–4579. [CrossRef] [PubMed]
10. Woese, C.R. On the evolution of cells. *Proc. Natl. Acad. Sci. USA* **2002**, *99*, 8742–8747. [CrossRef] [PubMed]
11. Haudiquet, M.; de Sousa, J.M.; Touchon, M.; Rocha, E.P.C. Selfish, promiscuous, and sometimes useful: How mobile genetic elements drive horizontal gene transfer in microbial populations. *Philos. Trans. R. Soc. B* **2022**, *377*, 20210234. [CrossRef] [PubMed]
12. Keen, E.C.; Bliskovsky, V.V.; Malagon, F.; Baker, J.D.; Prince, J.S.; Klaus, J.S.; Adhya, S.L.; Groisman, E.A. Novel "superspreader" bacteriophages promote horizontal gene transfer by transformation. *mBio* **2017**, *8*, e02115-16. [CrossRef] [PubMed]
13. Griffith, F. The significance of pneumococcal types. *J. Hyg.* **1928**, *27*, 113–159. [CrossRef] [PubMed]
14. Wilhelm, S.W.; Suttle, C.A. Viruses and nutrient cycles in the sea. *BioScience* **1999**, *49*, 781–788. [CrossRef]
15. Smith, D.; Palacios-Pérez, M.; Jheeta, S. The enclosed intestinal microbiome: Semiochemical signals from the Precambrian and their disruption by heavy metal pollution. *Life* **2022**, *12*, 287. [CrossRef] [PubMed]
16. Sagan, L. On the origin of mitosing cells. *J. Theor. Biol.* **1967**, *14*, 255–274. [CrossRef] [PubMed]
17. Dache, Z.A.A.; Thierry, A.R. Mitochondria-derived cell-to-cell communication. *Cell Reports* **2023**, *42*, 112728. [CrossRef] [PubMed]
18. Sabater, B. Evolution and function of the chloroplast. Current investigations and perspectives. *Int. J. Mol. Sci.* **2018**, *19*, 3095. [CrossRef] [PubMed]
19. Cornejo-Castillo, F.M.; Inomura, K.; Zehr, J.P.; Follows, M.J. Metabolic trade-offs constrain the cell size ratio in a nitrogen-fixing symbiosis. *Cell* **2024**, *187*, 1762–1768. [CrossRef] [PubMed]
20. Imachi, H.; Nobu, M.K.; Nakahara, N.; Morono, Y.; Ogawara, M.; Takaki, Y.; Takano, Y.; Uematsu, K.; Ikuta, T.; Ito, M.; et al. Isolation of an archaeon at the prokaryote-eukaryote interface. *Nature* **2020**, *577*, 519–525. [CrossRef] [PubMed]
21. Matta, S.K.; Rinkenberger, N.; Dunay, I.R.; Sibley, L.D. *Toxoplasma gondii* infection and its implications within the nervous system. *Nat. Rev. Microbiol.* **2021**, *19*, 467–480. [CrossRef] [PubMed]
22. Keeling, P.J.; McCutcheon, J.P. Endosymbiosis: The feeling is not mutual. *J. Theor. Biol.* **2017**, *434*, 75–79. [CrossRef] [PubMed]

23. McCutcheon, J.P. The genomics and cell biology of host-beneficial intracellular infections. *Annu. Rev. Cell Dev. Biol.* **2021**, *37*, 115–142. [CrossRef] [PubMed]
24. Donoghue, P.; Antcliff, J. Origins of multicellularity. *Nature* **2010**, *466*, 41–42. [CrossRef] [PubMed]
25. Ford, T.D. Pre-Cambrian fossils from Charnwood Forest. *Proc. Yorks. Geol. Soc.* **1958**, *31*, 211–217. [CrossRef]
26. Bobrovskiy, I.; Hope, J.M.; Ivantsov, A.; Nettersheim, B.J.; Hallman, C.; Brocks, J.J. Ancient steroids establish the Ediacaran fossil *Dickinsonia* as one of the earliest animals. *Science* **2018**, *361*, 1246–1249. [CrossRef] [PubMed]
27. Niklas, K.J.; Newman, S.A. The many roads to and from multicellularity. *J. Exp. Bot.* **2020**, *71*, 3247–3253. [CrossRef] [PubMed]
28. Calhoun, S.; Mondo, S.J.; Grigoriev, I.V. Yeasts and how they came to be. *Nat. Rev. Microbiol.* **2019**, *17*, 649. [CrossRef] [PubMed]
29. Kovács, R.; Majoros, L. Fungal quorum-sensing molecules: A review of their antifungal effect against Candida biofilms. *J. Fungi* **2020**, *6*, 99. [CrossRef] [PubMed]
30. Reid, C.R. Thoughts from the forest floor: A review of cognition in the slime mould *Physarum polycephalum*. *Anim. Cogn.* **2023**, *26*, 1783–1797. [CrossRef] [PubMed]
31. Shu, D.-G.; Luo, H.-L.; Conway-Morris, S.; Zhang, X.-L.; Hu, S.-X.; Chen, L.; Han, J.; Zhu, M.; Li, Y.; Chen, L.-Z. Lower Cambrian vertebrates from South China. *Nature* **1999**, *402*, 42–46. [CrossRef] [PubMed]
32. Miyashita, T.; Coates, M.L.; Farrar, R.; Larson, P.; Manning, P.; Wogelius, R.A.; Edwards, N.P.; Anné, J.; Bergmann, U.; Palmer, A.R.; et al. Hagfish from the Cretaceous Tethys Sea and a reconciliation of the morphological-molecular conflict in early vertebrate phylogeny. *Proc. Natl. Acad. Sci. USA* **2019**, *116*, 2146–2151. [CrossRef] [PubMed]
33. Glover, C.N.; Weinrauch, A.M. The good, the bad and the slimy: Experimental studies of hagfish digestive and nutritional physiology. *J. Exp. Biol.* **2019**, *222*, jeb190470. [CrossRef] [PubMed]
34. Nakashima, K.; Kimura, S.; Ogawa, Y.; Watanabe, S.; Soma, S.; Kaneko, T.; Yamada, L.; Sawada, H.; Tung, C.-H.; Lu, T.-M.; et al. Chitin-based barrier immunity and its loss predated mucus-colonization by indigenous gut microbiota. *Nat. Commun.* **2018**, *9*, 3402. [CrossRef] [PubMed]
35. Matsushita, M. The complement system of agnathans. *Front. Immunol.* **2018**, *9*, 1405. [CrossRef] [PubMed]
36. Stark, J.M.; Stewart, J.R.; Blackburn, D.G. Phylogeny and evolutionary history of the amniote egg. *J. Morphol.* **2021**, *282*, 1080–1122. [CrossRef] [PubMed]
37. Grigg, G.; Nowack, J.; Bicudo, J.E.P.W.; Bal, N.C.; Woodward, H.N.; Seymour, R.S. Whole-body endothermy: Ancient, homologous, and widespread among the ancestors of mammals, birds and crocodylians. *Biol. Rev. Camb. Philos. Soc.* **2022**, *97*, 766–801. [CrossRef] [PubMed]
38. Yan, Z.; Lei, Y.; Zhao, P.; Zhang, D.; Shen, J.; Zhang, G.; Wei, R.; Liu, H.; Liu, X.; He, Y.; et al. Natural mating ability is associated with gut microbiota composition and function in captive male giant pandas. *Ecol. Evol.* **2024**, *14*, e11189. [CrossRef] [PubMed]

39. Reese, A.T.; Chadaideh, K.S.; Diggins, C.E.; Schell, L.D.; Beckel, M.; Callahan, P.; Ryan, R.; Thompson, M.E.; Carmody, R.N. Effects of domestication on the gut microbiota parallel those of human industrialization. *eLife* **2021**, *10*, e60197. [CrossRef] [PubMed]
40. Burkitt, D.P. Some diseases characteristic of modern western civilization. *Br. Med. J.* **1973**, *1*, 274–278. [CrossRef] [PubMed]
41. Margulis, L. Symbiogenesis and symbionticism. In *Symbiosis as a Source of Evolutionary Innovation: Speciation and Morphogenesis*; Margulis, L., Fester, R., Eds.; MIT Press: Cambridge, MA, USA, 1991; pp. 49–92.
42. Guerrero, R.; Margulis, L.; Berlanga, M. Symbiogenesis: The holobiont as a unit of evolution. *Int. Microbiol.* **2013**, *16*, 133–143. [PubMed]
43. Yong, E. *I Contain Multitudes: The Microbes within Us and a Grander View of Life*; Penguin Random House LLC.: London, UK, 2016; ISBN 978-1-784-70017-1.
44. Suzuki, T.A.; Fitzstevens, J.L.; Schmidt, V.T.; Enav, H.; Huus, K.; Mbong, M.; Adegite, B.R.; Zinsou, J.F.; Esen, M.; Velavan, T.; et al. Codiversification of gut microbiota with humans. *Science* **2022**, *377*, 1328–1332. [CrossRef] [PubMed]
45. Mohammadzadeh, R.; Mahnert, A.; Duller, S.; Moissl-Eichinger, C. Archaeal key-residents within the human microbiome: Characteristics, interactions and involvement in health and disease. *Curr. Opin. Microbiol.* **2022**, *67*, 102146. [CrossRef] [PubMed]
46. Laforest-Lapointe, I.; Arrieta, M.-C. Microbial eukaryotes: A missing link in gut microbiome studies. *mSystems* **2018**, *3*, e00201–17. [CrossRef] [PubMed]
47. Schneider, T. The holobiont self: Understanding immunity in context. *Hist. Philos. Life Sci.* **2021**, *43*, 99. [CrossRef] [PubMed]
48. Druvefors, E.; Myrelid, P.; Andersson, R.E.; Landerholm, K. Female and male fertility after colectomy and reconstructive surgery in inflammatory bowel disease: A national cohort study from Sweden. *J. Crohn's Colitis* **2023**, *17*, 1631–1638. [CrossRef] [PubMed]
49. Ermolaeva, M.A.; Schumacher, B. Insights from the worm: The elegans model for innate immunity. *Semin. Immunol.* **2014**, *26*, 303–309. [CrossRef] [PubMed]
50. Jiang, H.; Wang, D. The microbial zoo in the C. elegans intestine: Bacteria, fungi and viruses. *Viruses* **2018**, *10*, 85. [CrossRef] [PubMed]
51. Berg, M.; Zhou, X.Y.; Shapira, M. Host-specific functional significance of Caenorhabditis gut commensals. *Front. Microbiol.* **2016**, *7*, 1622. [CrossRef] [PubMed]
52. Haçariz, O.; Viau, C.; Karimian, F.; Xia, J. The symbiotic relationship between *Caenorhabditis elegans* and members of its microbiome contributes to worm fitness and lifespan extension. *BMC Genom.* **2021**, *22*, 364. [CrossRef] [PubMed]
53. Banchereau, J.; Briere, F.; Caux, C.; Davoust, J.; Lebecque, S.; Liu, Y.-J.; Pulendran, B.; Palucka, K. Immunobiology of dendritic cells. *Annu. Rev. Immunol.* **2000**, *18*, 767–811. [CrossRef] [PubMed]
54. Delgado, M.-G.; Lennon-Duménil, A.-M. How cell migration helps immune cells. *Front. Cell Dev. Biol.* **2022**, *10*, 932472. [CrossRef] [PubMed]
55. Noel, C.; Dufemez, F.; Gerbod, G.; Egcomb, V.P.; Delgado-Viscogliosi, P.; Ho, L.-C.; Singh, M.; Wintjens, R.; Soglin, M.L.; Capron, M.; et al. Molecular phylogenies of Blastocystis isolates from different hosts: Implications for genetic diversity, identification of species, and zoonosis. *J. Clin. Microbiol.* **2005**, *43*, 348–355. [CrossRef] [PubMed]

56. Scanlan, P.D.; Stensvold, C.R.; Rajilic-Stojanovic, M.; Heilig, H.G.H.J.; De Vos, W.M.; O'Toole, P.W.; Cotter, P.D. The microbial eukaryote *Blastocystis* is a prevalent and diverse member of the healthy human gut microbiota. *FEMS Microbiol. Ecol.* **2014**, *90*, 326–330. [CrossRef] [PubMed]
57. Popruk, S.; Adao, D.E.V.; Rivera, W. Epidemiology and subtype distribution of *Blastocystis* in humans: A review. *Infect. Genet. Evol.* **2021**, *95*, 105085. [CrossRef] [PubMed]
58. Asnicar, F.; Berry, S.E.; Valdes, A.M.; Nguyen, L.H.; Piccinno, G.; Drew, D.A.; Leeming, E.; Gibson, R.; Le Roy, C.; Al Khatib, H.; et al. Microbiome connections with host metabolism and habitual diet from 1,098 deeply phenotyped individuals. *Nat. Med.* **2021**, *27*, 321–332. [CrossRef] [PubMed]
59. Lepczyńska, M.; Białkowska, J.; Dzika, E.; Piskorz-Ogórek, K.; Korycińska, J. *Blastocystis*: How do specific diets and human gut microbiota affect its development and pathogenicity? *Eur. J. Clin. Microbiol. Infect. Dis.* **2017**, *36*, 1531–1540. [CrossRef] [PubMed]
60. Escallón, C.; Belden, L.K.; Moore, I.T. The cloacal microbiome changes with the breeding season in a wild bird. *Integr. Org. Biol.* **2019**, *1*, oby009. [CrossRef] [PubMed]
61. Hird, S.M.; Carstens, B.C.; Cardiff, S.W.; Dittman, D.L.; Brumfield, R.T. Sampling locality is more detectable than taxonomy or ecology in the gut microbiota of the brood-parasitic Brown-headed Cowbird (*Molothrus ater*). *PeerJ* **2014**, *2*, e321. [CrossRef] [PubMed]
62. Ding, J.; Liao, N.; Zheng, Y.; Yang, L.; Zhou, H.; Xu, K.; Han, C.; Luo, H.; Qin, C.; Tang, C.; et al. The composition and function of pigeon milk microbiota transmitted from parent pigeons to squabs. *Front. Microbiol.* **2020**, *11*, 1789. [CrossRef] [PubMed]
63. Kim, S.Y.; Yi, D.Y. Analysis of the human breast milk microbiome and bacterial extracellular vesicles in healthy mothers. *Exp. Mol. Med.* **2020**, *52*, 1288–1297. [CrossRef] [PubMed]
64. Waddington, C.H. *Toward a Theoretical Biology*; The Basic Ideas of Biology; Edinburgh University Press: Edinburgh, UK, 1968; pp. 1–32.
65. Fitz-James, M.H.; Cavalli, G. Molecular mechanisms of transgenerational epigenetic inheritance. *Nat. Rev. Genet.* **2022**, *23*, 325–341. [CrossRef] [PubMed]
66. Jones, S.K.; McCarthy, D.M.; Vied, C.; Bhide, P.G. Transgenerational transmission of aspartame-induced anxiety and changes in glutamate-GABA signaling and gene expression in amygdala. *Proc. Natl. Acad. Sci. USA* **2022**, *119*, e2213120119. [CrossRef] [PubMed]
67. Horsthemke, B. A critical view on transgenerational epigenetic inheritance in humans. *Nat. Commun.* **2018**, *9*, 2973. [CrossRef] [PubMed]
68. Tun, H.M.; Chari, R.; Field, C.J.; Guttman, D.S.; Becker, A.B.; Mandhane, P.J.; Turvey, S.E.; Subbarao, P.; Sears, M.R.; Scott, J.A.; et al. Roles of birth mode and infant gut microbiota in intergenerational transmission of overweight and obesity from mother to offspring. *JAMA Pediatr.* **2018**, *172*, 368–377. [CrossRef] [PubMed]
69. Qin, Y.; Wade, P.A. Crosstalk between the microbiome and the epigenome: Messages from bugs. *J. Biochem.* **2018**, *163*, 105–112. [CrossRef] [PubMed]
70. Xu, X.; Miao, Z.; Sun, M.; Wan, B. Epigenetic mechanisms of paternal stress in offspring development and diseases. *Int. J. Genom.* **2021**, *2021*, 6632719. [CrossRef] [PubMed]

71. Belvoncikova, P.; Splichalova, P.; Videnska, P.; Gardlik, R. The human mycobiome: Colonization, composition and the role in health and disease. *J. Fungi* **2022**, *8*, 1046. [CrossRef] [PubMed]
72. Zhang, F.; Aschenbrenner, D.; Yoo, J.Y.; Zou, T. The gut mycobiome in health, disease, and clinical applications in association with the gut bacterial microbiome assembly. *Lancet* **2022**, *3*, E969–E983. [CrossRef] [PubMed]
73. Ruotsalainen, A.L.; Tejesvi, M.V.; Vänni, P.; Suokas, M.; Tossavainen, P.; Pirttilä, A.M.; Talvensaari-Mattila, A.; Nissi, R. Child type 1 diabetes associated with mother vaginal bacteriome and mycobiome. *Med. Microbiol. Immunol.* **2022**, *211*, 185–194. [CrossRef] [PubMed]
74. Guo, W.; Liu, T.; Neves, A.L.A.; Long, R.; Degen, A.; Zhou, M.; Chen, X. Transmission of fungi and protozoa under grazing conditions from lactating yaks to suckling yak calves in early life. *Appl. Microbiol. Biotechnol.* **2023**, *107*, 4931–4945. [CrossRef] [PubMed]
75. Quagliariello, A.; Modi, A.; Innocenti, G.; Zaro, V.; Barbaro, C.C.; Ronchitelli, A.; Boschin, F.; Cavazzuti, C.; Dellù, E.; Radina, F.; et al. Ancient oral microbiomes support gradual Neolithic dietary shifts towards agriculture. *Nat. Commun.* **2022**, *13*, 6927. [CrossRef] [PubMed]
76. Gancz, A.S.; Farrer, A.G.; Nixon, M.P.; Wright, S.; Arriola, L.; Adner, C.; Davenport, E.R.; Gully, N.; Copper, A.; Britton, K.; et al. Ancient dental calculus reveals oral microbiome shifts associated with lifestyle and disease in Great Britain. *Nat. Microbiol.* **2023**, *8*, 2315–2325. [CrossRef] [PubMed]
77. Khalid, M.; Abdollahi, M. Role of lead in dental diseases. *J. Environ. Sci. Health* **2020**, *38*, 329–361. [CrossRef] [PubMed]
78. Wild, C.P. Complementing the genome with an "exposome": The outstanding challenge of environmental exposure measurement in molecular epidemiology. *Cancer Epidemiol. Biomark.* **2005**, *14*, 1847–1850. [CrossRef] [PubMed]
79. Canali, S. What is new about the exposome? Exploring scientific change in contemporary epidemiology. *Int. J. Environ. Res. Public Health* **2020**, *17*, 2879. [CrossRef] [PubMed]
80. Peters, A.; Nawrot, T.S.; Baccarelli, A.A. Hallmarks of environmental insults. *Cell* **2021**, *184*, 1455–1468. [CrossRef] [PubMed]
81. Halden, R.U. Plastics and health risks. *Annu. Rev. Public Health* **2010**, *31*, 179–194. [CrossRef] [PubMed]
82. Trasande, L.; Krithivasan, R.; Park, K.; Obsekov, V.; Belliveau, M. Chemicals used in plastic materials: An estimate of the attributable disease burden and costs in the United States. *J. Endocr. Soc.* **2024**, *8*, bvad163. [CrossRef] [PubMed]
83. McConnell, J.R.; Wilson, A.I.; Stohl, A.; Arienzo, M.M.; Chellman, N.J.; Eckhardt, S.; Thompson, E.M.; Pollard, A.M.; Steffensen, J.P. Lead pollution recorded in Greenland ice indicates European emissions tracked plagues, wars and imperial expansion during antiquity. *Proc. Natl. Acad. Sci. USA* **2018**, *115*, 5726–5731. [CrossRef] [PubMed]
84. Barbante, C.; Veysseyre, A.; Ferrari, C.; van der Velde, C.M.; Capodaglio, G.; Cescon, P.; Scarponi, G.; Boutron, C. Greenland snow evidence of large scale atmospheric contamination for platinum, palladium and rhodium. *Environ. Sci. Technol.* **2001**, *35*, 835–839. [CrossRef] [PubMed]

85. Sonnenburg, E.D.; Smits, S.A.; Tikhonov, M.; Higginbottom, S.A.; Wingreen, N.S.; Sonnenburg, J.L. Diet-induced extinctions in the gut microbiota compound over generations. *Nature* **2016**, *529*, 212–215. [CrossRef] [PubMed]
86. Carter, M.M.; Olm, M.R.; Merrill, B.D.; Dahan, D.; Tripathi, S.; Spencer, S.P.; Yu, F.B.; Jain, S.; Neff, N.; Jha, A.R.; et al. Ultra-deep sequencing of Hadza hunter-gatherers recovers vanishing gut microbes. *Cell* **2023**, *186*, 3111–3124. [CrossRef] [PubMed]
87. Shao, M.; Zhu, Y. Long term metal exposure changes gut microbiota of residents surrounding a mining and smelting area. *Sci. Rep.* **2020**, *10*, 4453. [CrossRef] [PubMed]
88. Dolfini, A. From the Neolithic to the Bronze Age in Central Italy: Settlement, burial, and social change at the dawn of metal production. *J. Archaeol. Res.* **2020**, *28*, 503–556. [CrossRef]
89. Kramberger, B.; Berthold, C.; Spiteri, C. Fifth millennium BC miniature ceramic bottles from the south-eastern Prealps and Central Balkans: A multi-disciplinary approach to study their content and function. *J. Archaeol. Sci. Rep.* **2021**, *38*, 102993. [CrossRef]
90. Dixson, A.F.; Dixson, B.J. Venus figurines of the early paleolithic: Symbols of fertility or attractiveness? *J. Anthropol.* **2011**, *2011*, 569120. [CrossRef]
91. Thompson, R.C.; Allam, A.H.; Lombardi, G.G.; Wann, L.S.; Sutherland, M.L.; Sutherland, J.D.; Soliman, M.A.; Frohlich, B.; Mininberg, D.T.; Monge, J.M.; et al. Atherosclerosis across 4000 years of human history: The Horus study of four ancient populations. *Lancet* **2013**, *381*, 1211–1222. [CrossRef] [PubMed]
92. Gasbarrini, G.; Rickards, O.; Martinez-Labarga, C.; Pacciani, E.; Chilleri, F.; Laterza, L.; Marangi, G.; Scaldaferri, F.; Gasbarrini, A. Origin of celiac disease: How old are predisposing haplotypes? *World J. Gastroenterol.* **2012**, *18*, 5300–5304. [CrossRef] [PubMed]
93. McMullen, R.L.; Dell'Acqua, G. History of natural ingredients in cosmetics. *Cosmetics* **2023**, *10*, 71. [CrossRef]
94. Bostock, J. Case of a periodical affection of the eyes and chest. *Med. Chir. Trans.* **1819**, *10*, 161–165. [CrossRef] [PubMed]
95. Conteville, L.C.; Oliveira-Ferreira, J.; Vicente, A.C.P. Heavy metal resistance in the Yanomami and Tunapco microbiome. *Mem. Inst. Oswaldo Cruz.* **2023**, *118*, e230086. [CrossRef] [PubMed]
96. Hof, A.R.; Bright, P.W. Quantifying the long-term decline of the West European Hedgehog in England by sub-sampling citizen science datasets. *Eur. J. Wildl. Res.* **2016**, *62*, 407–413. [CrossRef]
97. Wagner, D.L.; Grames, E.M.; Forister, M.L.; Berenbaum, M.R.; Stopak, D. Insect decline in the Anthropocene: Death by a thousand cuts. *Proc. Natl. Acad. Sci. USA* **2021**, *118*, e2023989118. [CrossRef] [PubMed]
98. Slobodian, M.R.; Petahtegoose, J.D.; Wallis, A.L.; Levesque, D.C.; Merritt, T.J.S. The effects of essential and non-essential metal toxicity in the *Drosophila melanogaster* insect model: A review. *Toxics* **2021**, *9*, 269. [CrossRef] [PubMed]
99. Bala, S.; Garg, D.; Thirumalesh, B.V.; Sharma, M.; Sridhar, K.; Inbaraj, B.S.; Tripathi, M. Recent strategies for bioremediation of emerging pollutants: A review for a green and sustainable environment. *Toxics* **2022**, *10*, 484. [CrossRef] [PubMed]

100. Ayilara, M.S.; Babalola, O.O. Bioremediation of environmental wastes: The role of microorganisms. *Front. Agron.* **2023**, *5*, 1183691. [CrossRef]
101. Bist, P.; Choudhary, S. Impact of heavy metal toxicity on the gut microbiota and its relationship with metabolites and future probiotics strategy: A review. *Biol. Trace Elem. Res.* **2022**, *200*, 5328–5350. [CrossRef] [PubMed]

6. The Primacy of the Immune System: A Dual Inheritance

6.1. Identification—Friend or Foe?

It seems reasonable to suppose that the single major contribution that a microbiome, "our own microbes," can make to its animal host is by strengthening its immune system. Clearly, we live in a world of opportunistic scavengers, invisible microbes as well as visible predators, and accurate defences are needed from the moment of birth. The emphasis in this chapter is on how the dual inheritance system ought to work, and what makes it go wrong, leaving subsequent chapters for a more specific discussion of individual conditions and, hopefully, their amelioration. In essence, while the child is developing in the womb according to its parental genetic inheritance, (thus far hypothetical) microbial sentinel cells exist in the microbiome of the mother. At birth, these cells have evolved to carry information about the up-to-date microbial environment of the local population, and are passed on in a "second inheritance" (i.e., supplementary to the "first" parental genetic inheritance) in order to activate the immune cells of the newborn child. In this way, information is provided to distinguish friend from foe, preventing own-body autoimmune reactions, as well as reactivity against the fundamentally harmless antigens found in peanuts or pollen grains, for example. In addition to this "calibration" of the neonate immune system (i.e., calibration against the microbial environment experienced by the mother), the microbiome has a second, related role: to stimulate growth of the gut–brain axis into adulthood by its control over the rate of peristalsis. Any failure to transmit the microbiome during the birth process increases the propensity for the development of the previously mentioned "triple plague" of the immune system, excess weight gain (i.e., the gut), and poor mental health (the brain). Interestingly, it also seems reasonable to suppose that subsequent microbes may encounter resistance from the already established immune system of the adult, with implications for the mode of action of probiotics, for example.

6.2. Mother's Milk

Not surprisingly, perhaps, breast milk is precisely formulated to support the infant [1], although preterm infants require more than usual care [2], and a milk bank should be available in case the mother is unable to express it herself [3]. It is interesting to note that microbes, including the supposedly "good" bifidobacteria, are present in breast milk [4], although there is little evidence for their actual incorporation into the infant microbiome [5]. Indeed, it may be that, rather than having a specific function, these relatively harmless entities are tolerated within the milk ducts specifically in order to stimulate the immune system of the baby. Alongside such microbes, breast milk contains galactose-based oligosaccharides, a form of dietary fibre that cannot readily be digested by our own enzymes and that, therefore, represents a system that has evolved specifically to feed the intestinal microbiome [1].

6.3. Handshaking: A Computer Analogy

As indicated in Chapter 2, it seems that a computer virus is more or less a direct analogy of its biological equivalent, replicating itself in the operating system of the computer/cell. Likewise, in computing terminology, the term *handshaking* refers to the initial exchange of the protocols of communication between two devices, and something similar must happen in the neonate, as the maternal microbiome interacts with the developing gut wall. Accordingly, while the parental genes sculpt the immature brain, gut wall and initially naïve immune system, after birth, the maternal microbial inheritance both calibrates the immune system against the microbial environment of the mother and activates the gut wall to initiate the formation of the gut–brain axis (Figure 2, Chapter 2). Following these necessary "handshaking" steps, in principle, the brain and gut wall fully develop their communication system to guide the growth of the child through to adulthood. It is important to note that the fully functioning intestinal microbiome is best considered as a single entity, a microbial community in a symbiotic relationship with its host as a holobiont. In particular, a successful handshaking process leads to the formation of working semiochemical receptors, without which gut–brain communication cannot take place. It is likely that the appearance of non-communicable disease corresponds to the microbiome becoming a disconnected collection of microbes, with consequent problems for the functioning of the gut–brain axis, at least partly through the vagal nerves. In turn, the absence of a fully functioning gut–brain axis interferes with the growth of the brain, potentially changing the relative development of its different parts. To emphasise the point, the gut contents have been associated with depression, probably signalling directly to the vagus nerve through the recently discovered neuropod cells [6].

6.4. Semiochemicals

The term "semiochemical" implies a signalling molecule acting between individuals of either the same or different species. Needless to say, there are many different versions of the term, of which perhaps the most well-known is the pheromone, meaning a volatile molecule acting as a sex attractant between individuals of the same species. Although mostly associated with insects, semiochemicals are also known to affect the behaviour of mammals such as dogs and pigs [7]. In this book, the word is used to refer to communication between the vertebrate body on the one hand, and its microbiome partner on the other. It could be that the molecules produced by the intestinal microbiota can be divided into two classes: chemicals produced for local action on the one hand, and semiochemicals, compounds designed to invoke a longer-range response, on the other. The cells lining the intestine are obviously well placed to benefit from bacterial metabolites produced within the intestinal lumen, and it has been observed that production of the hormone serotonin is facilitated by the effect of short-chain fatty acids (SFCAs) on enterochromaffin cells [8]. Likewise, bacteria in the colon supply the body with at least some of its vitamins [9]. Thus, dopamine, although originally described as a neurotransmitter, is speculated to be involved in bacteria–body signalling when found in the intestinal lumen [10]. It has even been suggested that the genes involved in the biosynthesis of these signalling molecules have been conserved since their initial development within the prokaryotes, being passed on by vertical gene transfer to plants, fungi, and animals at different times throughout their evolution [11]. Indeed, the evolution of cell-signalling molecules remains a subject of great interest, and of much speculation [12]. Similarly, recent observations surrounding neurological disorders such as Parkinson's and Alzheimer's have focused on specified "good" bacteria and the production of SCFAs on the one hand [13] and neurotransmitters on the other [14]. While much long-range signalling presumably takes place via the gut wall, at least some passes through the above-mentioned neuropod cells, sampling the intestinal lumen and directly informing the brain through the vagal nerves [15].

6.5. A Viral Shunt Mechanism

It is important to note that, while semiochemicals are produced by prokaryotes, especially bacteria, the genes that control their production can be passed on among many different groups. While the basic principle of horizonal gene transfer is that "transforming principle" stumbled upon by Frederick Griffith in the 1920s [16], the spread of antibiotic resistance has given new impetus to the field. Although such genes can be passed on in a variety of ways, perhaps the most direct is the disassembly of microbes by their corresponding phage viruses [17], followed by reuptake into new organisms [18]. The situation is remarkably reminiscent of what has been described as a viral shunt, in which nutrients, including carbon atom-containing molecules or ammonium ions, can be held within a layer of sea water by the phage-stimulated continuous destruction and reassembly of microbes [19]. In case it be thought that the microbiome is not quite like sea water, essentially the same processes have also been described in soil micro-compartments [20], perhaps more analogous to the situation found in the gut. This microbiome viral shunt mechanism not only explains the speed with which the bacterial composition of the microbiome may change, as described by Figure 5 (Chapter 5); it also provides a mechanism by which desirable mobile genetic functionality may be retained. As an example of this process, the ability of the microbiome to metabolise a seaweed-specific carbohydrate was found to have transferred itself from the nori-eating Japanese people, in which it was originally found, to the microbiome of seemingly unconnected people living in parts of America [21]. In essence, it appears that a key feature of the diverse bacteria of the fully functioning microbiome is the ability to delocalise function—making it independent of the precise bacterial composition. There is a degree of overlap with antibiotics here, in the sense that bacteriophage action in infected tissues is accompanied by immune and debris-removing lymphatic systems, thus pre-empting any re-emergence of reconstituted microbes via an antibiotic equivalent of the viral shunt mechanism [19]. The existence of this mechanism probably explains why it is difficult to find specific "probiotic" bacteria that were introduced in food and, for example, even if the bacteria present in breast milk evade the neonate immune system, they are likely to be recycled into other constituents by phage action [5]. Nevertheless, mobile gene functionality could be introduced into the neonate microbiome. In other words, even if the microbes themselves are transitory, that does not stop them from being valuable.

6.6. *An Uncertainty Microbiome?*

Following the discussion above, when the microbiome is viewed from the point of view of phages and the viral shunt mechanism [19], it is possible to envisage a Heisenberg-like uncertainty principle. As illustrated by Figure 5 (Chapter 5), this vision entails a soup of slowly disintegrating/emerging microbes existing in shifting, constantly changing equilibria, which only resolves itself into a precise composition (lactobacilli, bifidobacteria, etc.) when sampled. In this context, it is interesting to consider the fate of relatively innocuous food-borne external bacteria, potentially including probiotics. Assuming they avoid digestion and the immune system on the way down, it may not be long before they are "virally shunted" by phage viruses into the microbiome soup, rendering their mobile genetic elements available to be incorporated as functioning units [22], while the bacteria that express them are constantly being destroyed and reformed.

6.7. *An Officer Class?*

In this context, unicellular eukaryote benegens may well act as a "missing link," operating between the gut wall and bacterial microbiome [23]. As illustrated by Figure 6 (Chapter 5), and bearing in mind the ability of *Toxoplasma gondii* to effectively take control of the nervous system of a rodent [24], such microeukaryotes could act as control elements, influencing what genetic elements are taken up and expressed. In this way, the flexible microbiome allows the survival of both "halves" of the holobiont: the microbial community and its multicellular host, acting as a single unit. Conversely, it would be the loss of this unicellular eukaryote-coordinating ability that would give rise to non-communicable disease.

6.8. *Peristalsis and the Partition of Nutrition*

Granted that the primary role of the microbiome is to contribute to the immune system of the neonate, it will be necessary to ensure that it remains effective throughout periods of famine. The most efficient way to do this is probably to slow the rate of peristalsis (the flow of food through the digestive tract), as a slower rate will ensure greater absorption of nutrition into the host. Perhaps the most likely scenario is that, while the microbiome constantly deploys its semiochemicals to issue an insistent demand for food, accelerating peristalsis, famine will lead the host to counter their action with increased hormones. In essence, this would constitute a form of negotiation, carried out across the gut wall, in which the microbiome lowers its activity for the duration of the emergency. If the host survives the experience, an increase in the levels of nutrition will eventually reduce hormonal pressure on the gut wall, allowing normal semiochemical service to resume. In most circumstances, in the presence of a fully functioning microbiome, when the energy demands of the host are already satisfied, excess nutrition is taken up by high-diversity bacterial growth and eventual excretion [25], preventing obesity.

6.9. *Poor Gut–Brain Communication*

While the above paragraphs describe what ought to happen, unfortunately, due to widespread microbiome malfunction, nowadays this is rarely the case. Although experience and logic dictate that the primary result of microbiome failure at birth will be autoimmune disease, at this point it is more convenient to describe how such failure may also lead to gut disturbance and, possibly, eventual weight gain. Non-communicable disease is often associated with poor gut motility; probably because poor gut–brain connectivity leads to intermittent spasms in the intestine, as studied by the not-for-profit research group, the Rome Foundation [26]. Peristalsis is a complicated phenomenon, assessed either by modelling [27], or by experiments in laboratory animals [28], although it must be borne in mind that such animals have been found wanting in assessments of microbiome-related disease [29]. Indeed, it is possible that the complicated nature of peristalsis is largely due to the varying levels of microbiome malfunction in the populations that have been studied so far. In fact, it is the increased abdominal pressure upon defecation that gives rise to at least two of Burkitt's "diseases characteristic of modern western civilization" (Table 1, Chapter 2): hiatus hernia and varicose veins.

6.10. A Microbiome Energy Excretion Model (MEEM)

Hitherto the dominant description of weight gain has been the energy balance model (EBM), in which the "chemical potential energy" of food is converted into the "kinetic and thermal energy" associated with moving and keeping warm [30]. In essence, this model counts the excreted chemical potential energy contained within faeces as negligible, which is possibly a reasonable assumption based on the knowledge of people living in the conditions described by Denis Burkitt as Western Civilisation. It seems likely, however, that the people living in their traditional lifestyles, whether with high levels of dietary fibre or, like the Maasai, eating a less fibre-heavy, more Western-style diet, excrete non-negligible levels of faecal chemical potential energy [31]. This observation is backed up by data from 21st century sewage treatment works, illustrating a decreased weight of faecal output within "richer" societies, as assessed by the Human Development Index; see Figure 4 (Chapter 4) [32]. On the assumption that no vertebrate animal species has yet survived the widespread destruction of its microbiome partner, it seems likely that the set-point hypothesis of Keesey, mentioned earlier (Section 4.20), is no longer valid for "industrialised" humans [33]. Accordingly, human body weight and shape have been slowly changing over the years as semiochemical-controlled peristalsis, microbiome growth, and faecal energy excretion have failed. In turn, it is likely that this Microbiome Energy Excretion Model (MEEM) will ultimately explain the reduction in the average body temperature seen in American service personnel since the mid-19th century [34]. As detailed above (Section 4.20), attempts to lose weight by intensive exercise have been defeated by a process called energy compensation, in which subjects either eat more food or use less energy; for example, by a reduction in body temperature [35]. In essence, normal movement (technically, non-exercise activity thermogenesis) is treated as intensive exercise when carrying body mass significantly in excess of that expected for the species [36].

6.11. The Mathematics of Obesity

In a fully grown adult with a stable body weight, total energy output must equal the input received from digestible foods. In turn, such energy input is regulated by appetite-related hormonal control in order to ensure a species-related ideal weight for each individual, granted that semiochemical-controlled peristalsis, microbiome growth, and faecal energy excretion are unimpaired. Unlike the above-mentioned energy balance model [30], in which faecal energy excretion is considered to be negligible, the Microbiome Energy Excretion Model, or MEEM, can be expressed by Equation (1):

$$\text{Energy intake} = \text{Energy used (converted to carbon dioxide, } CO_2\text{)} + \text{Energy excreted (faeces)} \quad (1)$$

It can be seen that, given the above assumptions, eating more within the appetite range of the individual could lead to greater excretion while retaining body weight within an acceptable range for the species, bearing in mind potential regional variations. By contrast, the failure of semiochemical control, peristalsis, and microbe growth, in combination with low faecal energy excretion, would lead to an increase in body mass. Accordingly, final body mass would be due to a combination of factors: the level of appetite and proportion of digestible foods consumed, as well as the degree of damage to the microbiome (noting that so-called ultraprocessed foods tend to be highly digestible). Eventually, a new equilibrium will be reached, when a greater body mass results in greater use of energy (i.e., non-exercise activity thermogenesis [36]), but the resultant more-or-less stable, potentially obese, body mass may well lie outside the healthy range for the species. In this context, it is worth mentioning that farmed animals also increase their weight upon treatment with antibiotics [37], presumably by a similar mechanism. Of course, microbiome deficiency may affect the other half of the gut–brain axis and, consequently, poor mental health may also lead to disordered eating, resulting in either gain or loss of body weight (see Section 7.10, below).

6.12. Autoimmunity

As indicated in the previous chapter, in the absence of microbial sentinel cells, the "uncalibrated" neonate immune cells have not been directed to any specific target. Accordingly, the immune system may overreact to otherwise harmless substances: antigens in pollen, in food, or even potential antigens involved in normal bodily functions. Nevertheless, even the naïve immune system still has a job to do and, accordingly, the pattern of autoimmune disease described as atopic march can start very early, often within the first month of life [38]. Interestingly, these immune cells also react to pollen grains, producing the same immunoglobulin E and activation of granulocytes as if the body were attacking itself [39]. Following the initial interest in hay fever and asthma by David Strachan [40], food allergy is a relatively recent phenomenon that has been described as "the second wave of the allergy epidemic" [41]. A recent review on the pathogenesis of autoimmune disease blames "an interplay between a genetic predisposition and environmental factors," without specifying the precise nature of such an environment, but mentioning that "the microbiota can also influence pathogenesis" [42]. It is increasingly clear that systemic chronic inflammation is consistent with Barker's "Fetal and Infant Origins of Adult Disease" [43] in that it both starts in childhood and underlies many of the non-communicable diseases of adult life [44]. However, it could be that such chronic inflammation, and autoimmune disease, are just two sides of the same coin. In addition, as noted earlier, the dual inheritance hypothesis implies that the fully functioning microbiome carries important aspects of the environment directly to the neonate immune system, and its failure can work both ways, with overreaction on the one hand, and inability to eliminate precancerous cells on the other. In this way it could be giving rise to the observed increase in cancer victims aged 50 years or younger [45]. Perhaps the easiest way to envisage the loss of microbial sentinel cells is by the delivery of babies by caesarean section under sterile conditions.

6.13. Caesarean Section

It is clear that the early post-birth period is crucial for the development of both child and microbiome [46], a period sometimes taken as the first thousand days [47]. Perhaps the most obvious source of the failure to take up the maternal microbiome is delivery by caesarean section under sterile conditions, in which the initial uptake of microbiota is delayed, with consequences such as a poor response to vaccination, for example [48]. However, as infant infections are not increased on C-section delivery following maternal request [49], this procedure is likely to increase significantly. Unfortunately, there is no completely effective way to upgrade the microbiota of C-section-delivered babies. Perhaps the easiest approach is the use of a swab via so-called vaginal seeding, transferring microbes to the head of the baby, but with all the associated risks of infection by pathogens [50]. Interestingly, one attempt to avoid this problem is to add a theoretically "good" intestine-inhabiting microbe, *Bifidobacterium longum* subspecies *infantis*, as an "ecologically aware" approach to the problem [51]. While the consequences of this approach remain to be seen, one point to bear in mind is the likelihood that beneficial microbes are positively selected for, i.e., that relatively small amounts of such microbes may be unexpectedly successful, perhaps especially if they are unicellular eukaryote "benegens" rather than bacteria. On the other hand, if such microeukaryotes are lost from the maternal microbiome, they are unlikely to be replaced from the external environment.

6.14. Subclinical Lead Poisoning

Heavy metals are one type of pollutant that does pose a significant danger. Lead is a readily accessible element and has been a global source of pollution, especially following its use as an additive in gasoline up to the end of the 20th century [52]. It will have been spread as fine particles in dust and on food and is still detectable in the air of London, the capital city of the UK, even 20 years after its last use for this purpose [53]. Although lead exposure is measured by blood levels rather than its influence on the microbiome, the global health burden remains significant, especially concerning the IQ of children and its consequence for cardiovascular disease [54]. In addition, lead has been associated with the onset of attention-deficit hyperactivity disorder (ADHD) [55]. While many different types of microbes will be vulnerable to lead particle poisoning, eukaryotic microbial sentinel cells, with their remit to investigate previously unknown pathogen-like particles, would be more prone to being disabled. In turn, such disabled cells would be devoured by other components of the microbiome, thus liberating the metal ion to poison again. Accordingly, the consequences for the function of the microbiome and, hence, the individual holobiont adult, and its children and successors, would depend on the balance of a number of competing rates: the rates of metal contamination versus excretion on the one hand, and the regeneration of vulnerable microbes on the other. In this context, it is often noted that blood lead levels are associated with constipation, especially in children [56], but the relationship is not reliably predictive [57]. Note, however, that screening relates to the direct action of blood-measured lead on the nerves, rather than any direct effect on the microbiome itself. Overall, it is possible to imagine a sequence of events that led to Dr. Bostock's catarrhus aestivus [58]. For example, it could be that his mother was a regular user of 18th century-standard lead oxide-based lipstick [59], and that some particles entered her intestine at such a rate as to deplete her microbial sentinel cells over time. If so, her son may well have been born with the propensity for autoimmune disease. The association between hay fever and well-off people, likely users of such cosmetics, was noted by Dr. Bostock in a subsequent paper [60]. Although we can be sure that significant heavy metal contamination does change the bacteriome of affected species, including humans [61], it could be that the key change causing other aspects of non-communicable disease is the loss of the relatively neglected microeukaryotes themselves [23], whether or not the vulnerable entities were actually sentinel cells.

6.15. *Pewter, Mercury and the Dental Dilemma*

Essentially the same arguments apply to pewter, a workable tin-based alloy containing variable amounts of hardening agents, potentially including antimony and lead. Its extensive use for bowls, plates, and spoons will indeed have contributed to historical heavy metal poisoning, even after lead was removed because of its toxicity [62]. Similarly, while dental amalgam is normally considered to be safe on the grounds of low exposure to poisonous mercury vapour [63], the possibility of a deleterious effect of swallowed metal on the microbiome adds another angle to the vexed question of its potential toxicity. However, the fact that it was introduced in the mid-19th century means that it was not responsible for either the hay fever of John Bostock [58], or the early stirrings of the obesity epidemic as reported by Denis Burkitt [31]. Indeed, it could be that any toxic contribution to the microbiome from dental amalgam is lost in the overall burden of pollution accompanying industrialisation.

6.16. *Summary: The Critical Role of Peristalsis*

Professors Margulis and Woese helped launch revolutions in the way that we think about evolution, whose consequences are currently being worked through. While the admittedly unverifiable version of symbiosis presented in this chapter is not without its problems, it does provide a basis upon which to understand non-communicable disease: that the fundamental role of the microbiome is as an accessory to the vertebrate immune system.

Accordingly, the dual inheritance hypothesis can be expressed in the following ways, as previously illustrated by Figure 2 (Chapter 2):

1. That, when fully functioning, microeukaryotes somehow coordinate prokaryotes and their mobile genetic elements into a single community, a microbiome, and that this community acts within the intestine of a vertebrate animal, its primary purpose being to transfer information about the microbial environment of the mother to its offspring, be it egg or live birth.
2. That, in a vertebrate animal, the maternal microbiome contains microbial sentinel cells as the prime carriers of such information, and also agents that work with the gut wall to stimulate growth of the gut–brain axis.
3. That, in a full-grown vertebrate animal, the action of semiochemicals on the gut wall stimulates the gut–brain axis, in turn accelerating peristalsis and, therefore, the flow of nutrition to the microbiome (Figure 8). This action maintains it in readiness so that, in the female animal of child-bearing age, it may be passed on to the next generation during the birth process, whether by egg or by live birth. In the male, or female outside the fertile age range, the prime action of the microbiome is simply to stimulate peristalsis to avoid excess build-up of fat.

Note that the fully functioning, flexible microbiome coordinates with the body as a holobiont, coping with adverse circumstances such as famine, as illustrated in Figure 1 (Chapter 2). The failure of the microbiome at birth has multiple consequences, passing down the generations as follows:

1. The lack of coordinating microeukaryotes tends to make the microbiome act more like individual microbiota, disrupting the holobiont and compromising its immune system support function.
2. Prevention of the flow of nutrition to the microbiome by interrupting peristalsis increases absorption, tending to increase the weight of the individual and inducing irritable bowel syndrome.
3. A malfunctioning maternal microbiome is passed on as the inheritance of the next generation, ensuring the progression of non-communicable disease.

While alterations to the bacterial microbiome appear to affect the nature of disease, for example with probiotics, in reality it could be an interaction with the immune system that is of primary importance, as described in Figure 2 (Chapter 2). Part 3, therefore, deals with the way that the inheritance of microbiome deficiency affects our health.

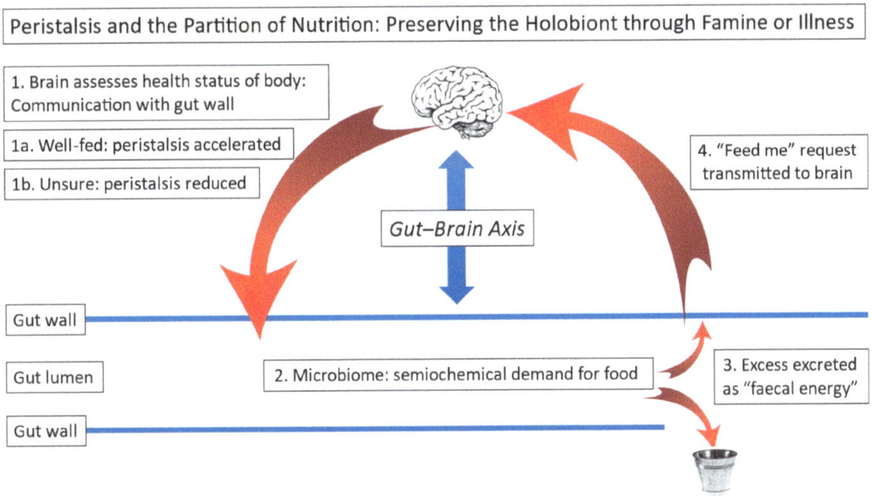

Figure 8. Peristalsis and the partition of nutrition. Box 1. It seems reasonable to suppose that the healthy holobiont allows constant communication along what has been termed the "microbiota–gut–brain axis". Box 1a: If both brain and gut report a healthy body, the acceleration of peristalsis ensures that the microbiome remains well fed. Box 1b: In famine, or if communication is interrupted for any reason, the brain has to assume the worst, and slows peristalsis until health and communication is resumed. Box 2: In essence, the microbiome uses messenger chemicals, semiochemicals, to request food from the body, albeit that this message may be modified throughout the various stages of pregnancy. Box 3: Of course, defecation in humans is a socially charged act, requiring delay until an appropriate level of privacy can be assured. Accordingly, final approval must come from the brain, through the gut–brain axis, before this latter stage of peristalsis can be accomplished. Box 4: This step draws attention to the initial "handshaking" processes, the exchange of the "protocols of communication", beginning immediately after birth and setting up the "immune–gut–brain triangle" illustrated in Figure 2 (Chapter 2). Source: Figure by the Author.

References

1. Porro, M.; Kundrotaite, E.; Mellor, D.D.; Munialo, C.D. A narrative review of the functional components of human breast milk and their potential to modulate the gut microbiome, the consideration of maternal and child characteristics, and confounders of breastfeeding, and their impact on risk of obesity later in life. *Nutr. Rev.* **2023**, *81*, 597–609. [PubMed]
2. Indrio, F.; Neu, J.; Pettoello-Mantovani, M.; Marchese, F.; Martini, S.; Salatto, A.; Aceti, A. Development of the gastrointestinal tract in newborns as a challenge for an appropriate nutrition: A narrative review. *Nutrients* **2022**, *14*, 1405. [CrossRef] [PubMed]

3. Sighaldeh, S.S.; Moridi, M.; Sabagh, V.G.; Kazemnejad, A.; Shateranni, F. Improving milk donation behaviour through an educational intervention based on the theory of planned behaviour: Study protocol of a cluster randomized controlled trial. *Int. Breastfeed. J.* **2023**, *18*, 53. [CrossRef] [PubMed]
4. Kim, S.Y.; Yi, D.Y. Analysis of the human breast milk microbiome and bacterial extracellular vesicles in healthy mothers. *Exp. Mol. Med.* **2020**, *52*, 1288–1297. [CrossRef] [PubMed]
5. Edwards, C.A.; Van Loo-Bouwman, C.A.; Van Diepen, J.A.; Schoemaker, M.H.; Ozanne, S.E.; Venema, K.; Stanton, C.; Marinello, V.; Rueda, R.; Flourakis, M.; et al. A systematic review of breast milk microbiota composition and the evidence for transfer to and colonisation of the infant gut. *Benef. Microbes* **2022**, *13*, 365–381. [CrossRef] [PubMed]
6. Mörkl, S.; Butler, M.I.; Wagner-Skacel, J. Gut-brain-crosstalk- the vagus nerve and the microbiota-gut-brain axis in depression. A narrative review. *J. Affect. Disord. Rep.* **2023**, *13*, 100607. [CrossRef]
7. McGlone, J.J.; Archer, C.; Henderson, M. Interpretive review: Semiochemicals in domestic pigs and dogs. *Front. Vet. Sci.* **2022**, *9*, 967980. [CrossRef] [PubMed]
8. Reigstad, C.S.; Salmonson, C.E.; Rainey, J.F., III; Szurszewski, J.H.; Linden, D.R.; Sonnenburg, J.L.; Farrugia, G.; Kashyap, P.C. Gut microbes promote colonic serotonin production through an effect of short-chain fatty acids on enterochromaffin cells. *FASEB J.* **2005**, *29*, 1395–1403. [CrossRef] [PubMed]
9. LeBlanc, J.G.; Chain, F.; Martin, R.; Bermùndez-Humarán, L.G.; Courau, S.; Langella, P. Beneficial effects on host energy metabolism of short-chain fatty acids and vitamins produced by commensal and probiotic bacteria. *Microb. Cell Factories* **2017**, *16*, 79. [CrossRef] [PubMed]
10. Sudo, N. Biogenic amines: Signals between commensal microbiota and gut physiology. *Front. Endocrinol.* **2019**, *10*, 504. [CrossRef] [PubMed]
11. Iyer, L.M.; Aravind, L.; Coon, S.L.; Klein, D.C.; Koonin, E.V. Evolution of cell-cell signaling in animals: Did late horizontal gene transfer from bacteria have a role? *Trends Genet.* **2004**, *20*, 292–299. [CrossRef] [PubMed]
12. Nair, A.; Chauhan, P.; Saha, B.; Kubatzky, K.F. Conceptual evolution of cell signaling. *Int. J. Mol. Sci.* **2019**, *20*, 3292. [CrossRef] [PubMed]
13. Li, Z.; Liang, H.; Hu, Y.; Lu, L.; Zheng, C.; Fan, Y.; Wu, B.; Zou, T.; Luo, X.; Zhang, X.; et al. Gut bacterial profiles in Parkinson's disease: A systematic review. *CNS Neurosci. Ther.* **2023**, *29*, 140–157. [CrossRef] [PubMed]
14. Chen, Y.; Xu, J.; Chen, Y. Regulation of neurotransmitters by the gut microbiota and effects on cognition in neurological disorders. *Nutrients* **2021**, *13*, 2099. [CrossRef] [PubMed]
15. Kaelberer, M.M.; Rupprecht, L.E.; Liu, W.W.; Weng, P.; Bohórquez, D.V. Neuropod cells: Emerging biology of the gut-brain sensory transduction. *Annu. Rev. Neurosci.* **2020**, *43*, 337–353. [CrossRef] [PubMed]
16. Griffith, F. The significance of pneumococcal types. *J. Hyg.* **1928**, *27*, 113–159. [CrossRef] [PubMed]

17. Lekunberri, I.; Subirats, J.; Borrego, C.M.; Balcazar, J.L. Exploring the contribution of bacteriophage to antibiotic resistance. *Environ. Pollut.* **2017**, *220 Pt B*, 981–984. [CrossRef] [PubMed]
18. Pfeifer, E.; Bonnin, R.E.; Rocha, E.R.C. Phage-plasmids spread antibiotic resistance genes through infection and lysogenic conversion. *mBio* **2022**, *13*, e01851-22. [CrossRef] [PubMed]
19. Wilhelm, S.W.; Suttle, C.A. Viruses and nutrient cycles in the sea. *BioScience* **1999**, *49*, 781–788. [CrossRef]
20. Kuzyakov, Y.; Mason-Jones, K. Viruses in soil: Nano-scale undead drivers of microbial life, biogeochemical turnover and ecosystem functions. *Soil. Biol. Biochem.* **2018**, *127*, 305–317. [CrossRef]
21. Hehemann, J.H.; Kelly, A.G.; Pudlo, N.A.; Martens, E.C.; Boraston, A.B. Bacteria of the human gut microbiome catabolize red seaweed glycans with carbohydrate-active enzyme updates from extrinsic microbes. *Proc. Natl. Acad. Sci. USA* **2012**, *109*, 19786–19791. [CrossRef] [PubMed]
22. Haudiquet, M.; de Sousa, J.M.; Touchon, M.; Rocha, E.P.C. Selfish, promiscuous, and sometimes useful: How mobile genetic elements drive horizontal gene transfer in microbial populations. *Philos. Trans. R. Soc. B* **2022**, *377*, 20210234. [CrossRef] [PubMed]
23. Laforest-Lapointe, I.; Arrieta, M.-C. Microbial eukaryotes: A missing link in gut microbiome studies. *mSystems* **2018**, *3*, e00201-17. [CrossRef] [PubMed]
24. Matta, S.K.; Rinkenberger, N.; Dunay, I.R.; Sibley, L.D. *Toxoplasma gondii* infection and its implications within the nervous system. *Nat. Rev. Microbiol.* **2021**, *19*, 467–480. [CrossRef] [PubMed]
25. Vandeputte, D.; Falony, G.; Veira-Silva, S.; Tito, R.; Joossens, M.; Raes, J. Stool consistency is strongly associated with gut microbiota richness and composition, enterotypes and bacterial growth rates. *Gut* **2016**, *65*, 57–62. [CrossRef] [PubMed]
26. Drossman, D.A. Functional gastrointestinal disorders: History, pathophysiology, clinical features and Rome IV. *Gastroenterology* **2016**, *150*, 1262–1279. [CrossRef] [PubMed]
27. Sensoy, I. A review on the food digestion in the digestive tract and the used in vitro models. *Curr. Res. Food Sci.* **2021**, *4*, 308–319. [CrossRef] [PubMed]
28. Sasidharan, A.; Peethambar, B.A.; Kumar, K.S.; Kumar, A.V.; Hiregange, A.; Fawkes, N.; Collins, J.F.; Grosche, A.; Vidyasagar, S. Advancing peristalsis deciphering in mouse small intestine by multi-parameter tracking. *Commun. Biol.* **2023**, *6*, 1237. [CrossRef] [PubMed]
29. Burberry, A.; Wells, M.F.; Limone, F.; Couto, A.; Smith, K.S.; Keaney, J.; Gillet, G.; van Gastel, N.; Wang, J.-Y.; Pietilainen, O.; et al. C9orf72 suppresses systemic and neural inflammation induced by gut bacteria. *Nature* **2020**, *582*, 89–94. [CrossRef] [PubMed]
30. Hall, K.D.; Heymsfield, S.B.; Kemnitz, J.W.; Klein, S.; Schoeller, D.A.; Speakman, J.R. Energy balance and its components: Implications for body weight regulation. *Am. J. Clin. Nutr.* **2012**, *95*, 989–994. [CrossRef] [PubMed]
31. Burkitt, D.P. Some diseases characteristic of modern western civilization. *Br. Med. J.* **1973**, *1*, 274–278. [CrossRef] [PubMed]

32. Rose, C.; Parker, A.; Jefferson, B.; Cartmell, E. The characterisation of feces and urine: A review of the literature to inform advanced treatment technology. *Crit. Rev. Environ. Sci. Technol.* **2015**, *45*, 1827–1879. [CrossRef] [PubMed]
33. Keesey, R.E.; Hirvonen, M.D. Body weight set-points: Determination and adjustment. *J. Nutr.* **1997**, *127*, 1875S–1883S. [CrossRef] [PubMed]
34. Protsiv, M.; Ley, C.; Lankester, J.; Hastie, T.; Parsonnet, J. Decreasing human body temperature in the United States since the Industrial Revolution. *eLife* **2020**, *9*, e49555. [CrossRef] [PubMed]
35. Careau, V.; Halsey, L.G.; Pontzer, H.; Ainslie, P.N.; Andersen, L.F.; Anderson, L.J.; Arab, L.; Baddou, I.; Bedu-Addo, K.; Blaak, E.E.; et al. Energy compensation and adiposity in humans. *Curr. Biol.* **2021**, *31*, 1–8. [CrossRef] [PubMed]
36. Levine, J.A. Non-exercise activity thermogenesis (Neat). *Nutr. Rev.* **2004**, *62*, S82–S97. [CrossRef] [PubMed]
37. Gaskins, H.R.; Collier, C.T.; Anderson, D.B. Antibiotics as growth promotants: Mode of action. *Anim. Biotechnol.* **2002**, *13*, 29–42. [CrossRef] [PubMed]
38. Yang, L.; Fu, J.; Zhou, Y. Research progress in atopic march. *Front. Immunol.* **2020**, *11*, 1907. [CrossRef] [PubMed]
39. Hill, D.A.; Spergel, J.M. The atopic march: Critical evidence and clinical relevance. *Ann. Allergy Asthma Immunol.* **2018**, *120*, 131–137. [CrossRef] [PubMed]
40. Strachan, D.P. Hay fever, hygiene and household size. *BMJ* **1989**, *299*, 1259–1260. [CrossRef] [PubMed]
41. Peters, R.L.; Krawiec, M.; Koplin, J.J.; Santos, A.F. Update on food allergy. *Pediatr. Allergy Immunol.* **2021**, *32*, 647–657. [CrossRef] [PubMed]
42. Pisetsky, D.S. Pathogenesis of autoimmune disease. *Nat. Rev. Nephrol.* **2023**, *19*, 509–524. [CrossRef] [PubMed]
43. Barker, D.J. The fetal and infant origins of adult disease. *BMJ* **1990**, *301*, 1111. [CrossRef] [PubMed]
44. Furman, D.; Campisi, J.; Verdin, E.; Carrera-Bastos, P.; Targ, S.; Franceschi, C.; Ferrucci, L.; Gilroy, D.W.; Fasano, A.; Miller, G.W.; et al. Chronic inflammation in the etiology of disease across the life span. *Nat. Med.* **2019**, *25*, 1822–1832. [CrossRef] [PubMed]
45. Ugai, T.; Sasamoto, N.; Lee, H.Y.; Ando, M.; Song, M.; Tamimi, R.M.; Kawachi, I.; Campbell, P.T.; Giovannucci, E.L.; Weiderpass, E.; et al. Is early-onset cancer an emerging global epidemic? Current evidence and future implications. *Nat. Rev. Clin. Oncol.* **2022**, *19*, 656–673. [CrossRef] [PubMed]
46. Mesa, D.M.; Loureiro, B.; Iglesia, I.; Gonzalez, S.F.; Olivé, E.L.; Algar, O.G.; Solana, M.J.; Cabero, M.J.; Sainz, T.; Martinez, L.; et al. The evolving microbiome from pregnancy to early infancy: A comprehensive review. *Nutrients* **2020**, *12*, 133. [CrossRef] [PubMed]
47. Robertson, R.C.; Manges, A.R.; Finlay, B.B.; Prendergast, A.J. The human microbiome and child growth—First 1000 days and beyond. *Trends Microbiol.* **2019**, *27*, 131–147. [CrossRef] [PubMed]
48. de Koff, E.M.; van Baarle, D.; van Houten, M.A.; Reyman, M.; Berbers, G.A.M.; van den Ham, F.; Chu, M.L.J.N.; Sanders, E.A.M.; Bogaert, D.; Fuentes, S. Mode of delivery modulates the intestinal microbiota and impacts the response to vaccination. *Nat. Commun.* **2022**, *13*, 6638. [CrossRef] [PubMed]

49. Guo, Y.; Murphy, M.S.Q.; Dimanlig-Cruz, S.; Walker, M.C.; Wen, S.W.; El-Chaâr, D. Infant infections following cesarean delivery on maternal request: A population-based cohort study. *J. Obstet. Gynaecol. Can.* **2024**, *46*, 102455. [CrossRef] [PubMed]
50. Ríos-Covian, D.; Langella, P.; Martín, R. From short- to long-term effects of C-section delivery on microbiome establishment and host health. *Microorganisms* **2021**, *9*, 2122. [CrossRef] [PubMed]
51. Duar, R.M.; Kyle, D.; Tribe, R.M. Reintroducing *B. infantis* to the cesarean-born neonate: An ecologically sound alternative to "vaginal seeding". *FEMS Microbiol. Lett.* **2020**, *367*, fnaa032. [CrossRef] [PubMed]
52. Needleman, H. The removal of lead from gasoline: Historical and personal reflections. *Environ. Res.* **2000**, *84*, 20–35. [CrossRef] [PubMed]
53. Resongles, E.; Dietze, V.; Green, D.C.; Harrison, R.M.; Ochoa-Gonzalez, R.; Tremper, A.H.; Weiss, D.J. Strong evidence for the continued contribution of lead deposited in the 20th century to the atmospheric environment of the London of today. *Proc. Natl. Acad. Sci. USA* **2021**, *118*, e21022791118. [CrossRef] [PubMed]
54. Larsen, B.; Sánchez-Triana, E. Global health burden and cost of lead exposure in children and adults: A health impact and economic modelling analysis. *Lancet Planet. Health* **2023**, *7*, E831–E840. [CrossRef] [PubMed]
55. Donzelli, G.; Carducci, A.; Llopis-Gonzalez, A.; Verani, M.; Llopis-Morales, A.; Cioni, L.; Morales-Suárez-Varela, M. The association between lead and attention-deficit/ hyperactivity disorder: A systematic review. *Environ. Res. Public Health* **2019**, *16*, 382. [CrossRef] [PubMed]
56. Zamani, N.; Hosseini, A.; Farnaghi, F.; Sayyari, A.; Gholami, N.; Imanzadeh, F.; Hadeiy, S.K.; Hajipour, M.; Salimi, A.; Philips, S.; et al. Blood lead level evaluation in children presenting with chronic constipation in Tehran-Iran: A cross-sectional study. *Sci. Rep.* **2023**, *13*, 2301. [CrossRef] [PubMed]
57. Raghu, V.K.; Nowalk, A.J.; Srinath, A.I. Should children with constipation undergo blood lead level screening? *Clin. Pediatr.* **2019**, *58*, 627–632. [CrossRef] [PubMed]
58. Bostock, J. Case of a periodical affection of the eyes and chest. *Med. Chir. Trans.* **1819**, *10*, 161–165. [CrossRef] [PubMed]
59. McMullen, R.L.; Dell'Acqua, G. History of natural ingredients in cosmetics. *Cosmetics* **2023**, *10*, 71. [CrossRef]
60. Bostock, J. Of the catarrhus aestivus or summer catarrh. *Med. Chir. Trans.* **1828**, *14*, 437–446. [CrossRef] [PubMed]
61. Shao, M.; Zhu, Y. Long term metal exposure changes gut microbiota of residents surrounding a mining and smelting area. *Sci. Rep.* **2020**, *10*, 4453. [CrossRef] [PubMed]
62. Hull, C. *Pewter*; Bloomsbury: New York City, NY, USA, 2008.
63. Dodes, J.E. The amalgam controversy: An evidence-based analysis. *JADA* **2001**, *132*, 348–356. [PubMed]

Part 3. The Consequences for People

The Propensity for Disease: An inadequate maternal microbial inheritance means that the person has an increased tendency for all the forms of non-communicable disease, which may be classed in order of visibility: obesity, with or without functional gastrointestinal disorders; autoimmune disease; and poor mental health. It is important to note, however, that the resultant mixture of conditions will vary in severity from barely detectable to fatal. Indeed, it is another variation of what may be called the Anna Karenina Principle, taken from Tolstoy's 1877 novel of the same name, which can be translated as follows: "All happy families (i.e. healthy people) are alike; each unhappy family (malfunctioning microbiome) is unhappy in its own way". We have seen already that a healthy person allied with an intact, fully functioning, microbiome *from birth* can be regarded as a healthy holobiont, while even an apparently healthy person, attached to a deficient microbiome, will possess their own, unique, "fingerprint" of the propensity for the development of non-communicable disease. As described below, there is a need for an analytical method to assess the status of each individual, particularly if that person will be in a position to transfer their microbiome to a newborn child. In turn, as described in Chapter 9, in this way we may be able to add key health-giving, presumably eukaryotic, "benegen" microbes to the child at the moment of birth, as well as giving support to those people already carrying the consequences of a poor microbial inheritance.

7. Non-Communicable Disease: The Triple Plague

7.1. An Early Start

I have written in the Preface about my own conditions: the development of the nuisance of hay fever; the dangerous (but reversible) gain of weight; and the sensory overload of my own, specific variety of autism-like developmental disease. This latter problem initially came with a silver lining, as the relative gain of intellectual ability enabled me to obtain an excellent chemistry degree, although the corresponding lack of social ability hampered my subsequent career. My autism was also associated with another nuisance: irritable bowel syndrome, a fact readily explained by the degradation of my gut–brain axis. We have seen already that disorders of gut motility are covered by the Rome Foundation [1], but the link with autism strongly suggests that its efforts should be combined with organisations covering psychiatric diseases, such as the American Psychiatric Association and their *Diagnostic and Statistical Manual of Mental Disorders* (known simply as the "DSM") [2]. While autoimmune diseases, such as asthma, may start early [3], even if disease is detected in later life, the propensity for the development of microbiome-related disorders starts within a short time of birth, and possibly within the first few hours, as suggested by Barker and his "fetal and infant origins" article [4]. Weight gain and poor mental health, and their consequences, may follow later in the child or adult, often as mixed conditions, such as type 2 diabetes and depression [5]. A non-comprehensive summary of various conditions and their characteristics are outlined below, bearing in mind that the full list of microbiome-related non-communicable diseases is yet to be written.

7.2. Weight Gain

We have seen that the steady gain of weight readily visible throughout the modern world [6], even in childhood [7], may well have its origins in the loss of microbiome semiochemical-mediated control of gut motility. However, the direct cause is the slowing of peristalsis, which allows more time for the digestion and consequent overabsorption of nutrition (Figure 8, Chapter 6). Accordingly, eventual obesity is rendered more likely by the extensive consumption of readily digested foods, be they defined as "fast" or "ultraprocessed" [8]. Following its use as an additive in petrol, the increased concentration of lead in the inner-city districts of Westernised countries [9], along with their easy access to relatively inexpensive, readily digested foods, may help to account for the prevalence of obesity in those areas. Although the use of leaded petrol has been discontinued [10], nevertheless its effects continue, as damage to the microbiome of the mother is reflected in problems with the neonate and, therefore, in its subsequent growth into a microbiome-compromised adult. Mining and the burning of coal have also added to the toxic load, presumably not only affecting the microbiome of current generations [11], but subsequent ones as well. Of course, it is the slow, multi-generational nature of weight gain that effectively normalises the situation across whole populations. Interestingly, as scientific understanding of the human body has increased over the past decades, so our bodies have been changing, an effect most clearly shown by the decrease in average body temperature over the last 150 years [12]. Likewise, because such diseases are latent, it is hard to tell if any individual is truly "healthy". Significantly, not only can obesity [13] be treated as a microbiome-related non-communicable disease but, possibly after the conversion of fat into cholesterol, so can ischaemic stroke [14], gallstones [15] and even the uric acid crystals of gout [16], the latter being associated with a wide range of cardiovascular diseases [17].

7.3. Autoimmunity and Cancer

While the immune system has been under active investigation for a long time, it is probably fair to say that it is still in the descriptive stage, whether across the whole of life [18], or in the initial interactions with its brand-new body [19]. Indeed, it is interesting to note that the study of the immune system has had autoimmunity at its heart [20], as if it were somehow at the middle of the system itself, rather than a previously unknown disease state first appearing mysteriously as Bostock's catarrhus aestivus in the early 19th century [21]. The so-called atopic march is an evolving set of conditions causing issues throughout childhood, often starting with babyhood eczema, but progressing to hay fever (allergic rhinitis) and the much more serious asthma [22] and food allergy [23]. Interestingly, there is an opposite side to autoimmunity, in the sense of a failure of the immune system to recognise and eliminate precancerous cells, the rogue elements of its own body. In turn, this failure helps to account for the current increase in cancer in the younger population [24].

7.4. Cancer: An Incipient Animal?

An important connection, which we should see more of in the future, is what might be called the gut–cancer axis. While pancreatic cancer normally kills rapidly, in rare cases slower growth can be observed, and these cases are accompanied by a diverse bacteriome, both within the intestine of the sufferer and, apparently, within the tumour itself [25]. Indeed, it is as if the tumour were a separate organ, or even a nascent animal in its own right. The search for tumour-related microbiomes is now well underway, and includes detailing the difference between colorectal cancer in "young", i.e., people below the age of 50 years, and the more usual subjects in which the age of onset is greater than 60 years [26]. Unfortunately, these articles all focus on bacteria, and do not mention the possible presence of microeukaryotes within the tumour, the search for which may represent a logical next step in our understanding of the development of cancer.

7.5. Coeliac Disease

As mentioned by Denis Burkitt [27], essentially all microbiome-related conditions are epitomised by coeliac disease: "an immune-mediated enteropathy against dietary gluten present in wheat, rye, and barley" involving "harmful gluten peptides" [28]. A review of the epidemiology of this disease suggests a worldwide increase, in both prevalence and incidence, across the last three decades. The author's rationale for this increase is the improved diagnosis and the globalisation of poor diets, alongside what have become the standard statements associated with changes in the microbiome: delivery of babies by caesarean section, the lack of breastfeeding, and the use of antibiotics (but without mentioning heavy metal pollution) [29]. Whatever the exact reason, however, the actual presentation of coeliac disease is enormously variable. Associated immune system overreaction can occur anywhere in the body, with or without nutrient malabsorption or, indeed, with no symptoms at all [28]. As with atopic march [22], it may be expressed differently in children and in adults [30] and is commonly observed in the extraintestinal manifestation of dermatitis herpetiformis [31]. Significantly, although coeliac disease shares some genetic characteristics with autoimmune-related type 1 diabetes [32], there is little genetic or epidemiological overlap with gut–brain axis-related type 2 diabetes [33], although both may exhibit similar bacterial microbiome changes. One recent article is a study of the presumably genetically susceptible first-degree relatives of adult sufferers [34]. The article looks for evidence of detoxification of the gluten peptides by breaking the normally metabolically stable glutamine–proline bond, but without success [34]. A second article attempted to study whether alteration in the gut bacteria and corresponding metabolome of infants can be used to predict the later onset of disease. Many such changes were, indeed, observed, the authors opining that they have established a road map for future studies to detect and prevent the development of autoimmunity [35]. Almost invariably, however, such papers actually attempt only to categorise the bacteriome, with an occasional passing reference to its microeukaryote or viral constituents, let alone mobile genetic elements. Significantly, the relationship between coeliac disease and psychiatric and neurological disorders has never been fully understood. It is easy to imagine a degree of sadness and anxiety caused by severe coeliac disease, but harder to comprehend the observed positive association with manic-depressive behaviour, schizophrenia, or bipolar disorder [36]. A non-causative comorbidity solves the problem and, indeed, coeliac disease is only a specific example of a general autoimmune/depression associative trend [37].

7.6. Rheumatism, Multiple Sclerosis, etc.

As illustrated in Table 1 (Chapter 2), alongside coeliac disease, thyrotoxicosis [38], rheumatoid arthritis [39], and multiple sclerosis [40] were mentioned by Denis Burkitt as being unheard of in people following traditional lifestyles [27]. Although all are primarily complex autoimmune diseases, at the present time, multiple sclerosis (MS) is, perhaps, the most interesting from the point of view of known microbial involvement, with a suggestion that the Epstein–Barr virus is, indeed, "a likely driver of inflammatory autoimmune disease" and that it "provides a target for future therapies" [41]. Equally, Guillain–Barré syndrome was first described in 1916 and is a demyelinating polyneuropathy that may develop after various infections, or even after immunisation against the possibility of infection. While well-recognised in richer countries, it constitutes an increasing danger in the rest of the world [42]. Finally, similar autoimmune diseases have been connected to antibodies against IgLON5, a cell surface protein involved in nerve cell connectivity [43].

7.7. Fatigue, Long COVID, and Other Mysteries

Although more formally known as SARS-CoV-2, (SARS standing for "severe acute respiratory syndrome") the novel coronavirus that sprang on the world became known as COVID-19 as a contraction of both "coronavirus disease" and "2019", the year this version was first described. While the direct effect of the virus itself varied from mild to deadly, some ex-patients began to notice that they were not getting better, communicating with one another in self-help groups. In the light of their efforts, the term "Long-haul COVID" or, more simply, "Long COVID" began to be coined, bringing their plight to the attention of healthcare professionals [44]. Although COVID itself was very much in the ascendant, and each individual exhibited very different symptoms, nevertheless it slowly became apparent that Long COVID resembled earlier mysteries, variously termed myalgic encephalomyelitis or chronic fatigue syndrome (ME/CFS), another set of potentially devastating long-haul conditions [45]. Of course, fatigue is an exasperating symptom, often not easily linked to a specific cause. Sporadic mentions of historical disease have been traced back to the period 1860 to 1910, remaining rare until the late 20th century [46] but essentially following the same trajectory as hay fever within microbiome-related autoimmune disease: present but rare in the 19th century, and common and of increasing severity in the 21st century [47]. Indeed, SARS-CoV-2 has been specifically described as "an instrumental trigger" of autoimmunity [48]. Among potentially many Long COVID-related individual complaints, chronic fatigue, orthostatic dizziness, and brain fog have been associated with autoimmune disease in a similar fashion to Post-Treatment Lyme Disease Syndrome [49]. Multisystem inflammatory syndrome remains a threat, both in the young [50] and in adults [51]. Thromboembolism is a problem in people with Long COVID, and is also related to the immune system by dysregulation of the complement system [52]. Nevertheless, the most mysterious characteristic of both Long COVID and ME/CFS is chronic fatigue. There could be a new form of autoimmune disease: an interdiction of the protein chain associated with mitochondrial ATP, either its generation or its utilisation [53], bearing in mind that there are many potentially targetable antigens involved in the mitochondrial energy-harvesting cascade, and each patient could present with a unique combination thereof. Overall, it seems that the recent onset of Long COVID has highlighted the full extent and complexity of microbiome-related diseases, and that each individual has a different mix of these triple plague conditions. Interestingly, it could be that such fatigue-related autoimmune disease has been slowly increasing in extent for many years, with only the most severe cases coming to the attention of healthcare professionals, at least until the advent of SARS-CoV-2. A recent perspective (February 2024) has highlighted the substantial remaining problems posed by Long COVID, especially the absence of validated treatment methods and the waning attention of governmental agencies as it moves down the political agenda. The article mentions microbiome dysbiosis, but without going into detail [54]. Finally, muscle damage has been recorded after exercise [55], and it could be that this effect is related to the muscle weakening observed in recent tranches of schoolchildren mentioned earlier [7].

7.8. Bowel, Skin, Brain, Mouth, Lung, etc.

This section illustrates the interrelatedness of the diseases that are probably due to microbiome-function deficiency. Bearing in mind the above arguments, it should be no surprise that the risk of inflammatory bowel disease is raised in patients exhibiting atopic dermatitis [56], but that "original investigation" article aimed to answer a direct question, and did not try to find any wider links with microbial communities, or with depression or cardiovascular disease, for example. Psoriatic arthritis is one of those probably microbiome-related conditions that, like coeliac disease, can express itself in a wide variety of ways, affecting the skin and bowel or causing eye inflammation (uveitis) or cardiovascular disease, among other issues [57]. The connection between acne and its unproven but suspect bacterium, *Propionibacterium acnes*, has already been mentioned [58], but earlier work on acne vulgaris seemed to establish an "axis" between gut, brain and skin that manifests itself alongside depression and anxiety [59]. Certainly, the concept of the intestinal microbiome as "first among equals" is entirely consistent with the dual inheritance hypothesis, due to its transfer by direct faecal "contamination" via either live birth or egg. Indeed, the entry of skin bacteria into the intestinal microbiome of the neonate following delivery of a baby by caesarean section seems to have unhelpful consequences [60]. Regarding the brain, Parkinson's disease has recently been widely recognised as a gut microbiome-related condition, with a wide variety of bacterial families associated with both positive and negative effects, but its relationship with the production of short-chain fatty acids is unclear, and possibly irrelevant [61]. Likewise, as part of the complex tangle that is Alzheimer's disease (AD), what has been called the "microbiota–gut-AD axis" has been subject to significant effort, albeit mostly involving the search for potentially important pro- or anti-inflammatory bacteria [62]. The same logic follows when looking at the neurotransmitter-related connection between gut bacteria and AD [63], and with inflammatory bowel disease [64]. Sticking with brain diseases, there is a connection between migraine and a similar set of inflammatory bowel diseases that are seen with AD [65]. Polycystic ovary syndrome is a little-understood but widespread disease, increasing in incidence, which is associated with anxiety, depression, cardiovascular disease, weight gain, and type 2 diabetes [66], in addition to bipolar disorder [67]. It is, therefore, somewhat surprising that a recent comprehensive review makes no mention of possible microbiome involvement [68], an oversight that cannot last long.

7.9. Autoimmunity: A Case of Mistaken Identity?

It seems that the above-mentioned "harmful gluten peptides" of coeliac disease [28] must be surprisingly toxic for what is essentially a foodstuff [69]. Indeed, it could be that, bearing in mind the example of the skeleton of a young lady with suspected coeliac disease found in first century Italy [70], perhaps there is a resemblance to the rare initial appearance of the now-common hay fever [21]. It seems likely that the uncalibrated, naïve immune system described in Figure 2 (Chapter 2) has tagged these not-so-harmful gluten peptide antigens as "dangerous unknowns". In this view, coeliac disease represents a form of internal hay fever, in essence a debilitating, potentially deadly form of food allergy. Coeliac disease has been shown to be associated with inflammatory bowel diseases such as Crohn's and ulcerative colitis [71], and there is an established relationship with the autoimmune thyroid diseases [72]. It is likely that all these stem from fundamentally the same problem: a microbiome-related propensity toward autoimmune disease in general. Accordingly, stimuli such as viral modification, as postulated for multiple sclerosis, may bring about more widespread autoimmune episodes. A recent article has drawn attention to a possible intermediary between viruses and autoimmune disease, as a study on the Zika virus identified "bystander" activation of CD8+ T cells causing neuropathology via "antigen-independent" cytotoxicity [73]. However, such stimuli could also be part of normal bodily changes and, accordingly, hay fever may start with puberty, or Grave's disease with pregnancy, to quote just two examples. Sadly, the autoimmune effect is unlikely to resolve itself upon removal of the trigger and, therefore, elimination of the Epstein–Barr virus is unlikely to prove a "miracle cure" for MS. Incidentally, while miracles themselves are, perhaps, not totally verified, the placebo effect described by the three-way interaction of the immune–gut–brain triangle may well be sufficiently powerful to deflect the autoimmune response itself, even if it cannot repair the damage already done.

7.10. Mental Health

While the neonate brain is primed for growth, the direction of that growth will depend on the input received and, accordingly, requires a degree of engagement with the developing gut–brain axis. While much available capacity will be taken up with the growth of the senses, there is also a need for internal maps: both of proprioception, to do with the relationship between the body and the outside world, and of interoception, the status of its various components [74]. A deficit in the gut–brain axis is likely to have its greatest impact on this latter facility, as illustrated by the above-mentioned connection between autism and joint hypermobility [75], probably due to distortion of growth following poor communication between body and brain. Once this principle is accepted, a number of psychological phenomena become more readily explicable: as an example, while the correct response to apparent food poisoning is the diarrhoea and vomiting that is all too frequently observed by people travelling far from home [76], the lack of information from the gut may prompt an anxious, uncertain response for which the expression "butterflies in the stomach" seems to be appropriate. Indeed, anxiety disorders are very common, and are found in a wide variety of situations, as described in a recent comprehensive review [77], and increasingly found alongside depression and insomnia [78]. Significantly, as described above, the seemingly pointless rise in blood pressure known as white coat hypertension [79] is, presumably, also a form of anxiety disorder. In a similar vein, anorexia nervosa is recognised as the deadliest of the psychiatric disorders and, although a recent comprehensive review has described "interesting" intestinal microbiome characteristics, it was stressed that it is unknown whether these are a cause or consequence of the disease [80]. Of course, even young patients are only assessed when their neurodevelopmental conditions are well advanced, often presenting with many comorbidities, rather than being able to follow any distortion of mental image consequent upon the postnatal development of a weak gut–brain axis. Although microbial involvement has not been mentioned, the recently declared "epidemic" of loneliness [81] may well be related to an autism-like inability to communicate in socially significant environments. Similar developmental disorder arguments could be applied to many such conditions, including body mass dissatisfaction and depression [82]. Likewise, disordered eating [83], or even outright anorexia, may follow from a "too fat" distortion of reality [80]. Similarly, gender dysphoria adds a question mark over the very nature of disease itself [84], and is a source of significant professional dispute [85], while body dysmorphic disorder is, perhaps, not quite as contentious [86]. It is interesting to speculate that some of the difficulty in getting giant pandas to breed in captivity may be due to a disturbed microbiome [87], with potential consequences for their gut–brain axis and, possibly, neurodevelopmental disease. If so, this could be another example of "domestication" being equivalent to "industrialisation" in the microbiome of animals [88]. Perhaps the best way to understand the wide range of body dissatisfaction disorders is that the failure of the gut–brain axis in the neonate leads to distorted interoception on subsequent growth, which comes to the fore only when the individual is old enough to appreciate (if not fully understand) that their physical mirror reflection does not agree with their virtual self-image.

7.11. Brain Shape

Recent research has been uncovering the relationship between the distortion of brain shape and psychiatric conditions: striatal volume with psychopathy [89], evidence of brain asymmetry changes in autism [90], or in depression [91]. Likewise, both inflammation and changes in brain structure with neuropsychiatric disorders such as schizophrenia have been recorded [92], while the immune system has also been linked to autism via inappropriate neuroactivity and the gut microbiota [93]. However, as mentioned earlier, Mendelian randomisation studies suggest that, although poor mental health and inflammation are linked, consistent with the dual inheritance hypothesis, it seems that one does not actually cause the other [94]. Although autism-like disease can be hard to diagnose, its prevalence is clearly increasing in step with the autoimmune and weight gain epidemics of the modern age [95], and it is also reported to be strongly heritable [96]. Finally, gut microbiome studies show no particular commonality among sufferers of either autism or attention-deficit hyperactivity disorder [97], perhaps consistent with the idea that the precise bacterial (or mobile genetic element) composition of the intestinal microbiota is no longer relevant once the condition has been initiated from birth. In addition, once it is clear that both gut and brain disturbances are two sides of the same coin, i.e., the gut–brain axis, it should not be a surprise that type 2 diabetes and poor mental health are found to be related [7]. Equally, it should be no surprise that intermittent fasting causes both weight loss and changes in brain activity, as well as altering the relative proportions of intestinal bacteria [98], although the exact meaning of these changes is not so clear. Granted that there is a connection between brain shape on the one hand, and poor mental health on the other, it is possible that the weight gain normally associated with type 2 diabetes is also related to altered brain shape. Indeed, it could be that "normal" brain shape can only be found in those populations that do not suffer from the triple plagues of non-communicable disease. It is interesting to note an increase in brain volume between people born in the 1930s compared to those in the 1970s. The authors of the paper link this fact to the reduction in the incidence of dementia in recent years, largely on the basis of a corresponding increase in cognitive reserve, pointing out that early life environmental effects are a more likely cause than genetic changes, but without any mention of possible microbial involvement or, indeed, of the general increase in body mass index over this period [99].

7.12. Placebo Effect

While there is significant interest in the connection between nutrition and mental health [100], any improvement could be due to a strengthening of the gut–brain axis. Similarly, the appearance and disappearance of the symptoms of autoimmune disease may seem to be entirely random. However, as illustrated by the "triangle diagram" of Figure 2 (Chapter 2), essentially any input affects the rest, and the involvement of the brain sets up a resemblance to the placebo effect [101]. The system works both ways, however, and what is known as the nocebo effect may cause harm in certain circumstances [102]. Indeed, it has been referred to as modern voodoo, in that unconscious offhand treatment of an already ill patient has been suggested to actually accelerate death [103]. Of course, placebo effects are notoriously hard to study, being subject to all kinds of bias [104]. Interestingly, when studies were carried out without deception—stating specifically that "this is a sugar pill", for instance—the placebo effect still led to an improvement compared to no treatment [105]. The immune–gut–brain triangle illustrated by Figure 2 suggests a reason for the ubiquitous nature of this effect: that even a transitory disturbance in any one of the corners will affect the others. A doctor with a good "bedside manner", however that is defined, is likely to achieve better results than less gifted doctors and, while a good cup of tea may work wonders, it is possible that probiotics, whether patented or sauerkraut-like natural, improve apparent health by initially engaging with the immune system of the patient. As an example, since psychiatric conditions are associated with presumably irreversible alteration of brain structure, any changes in the symptoms of autism spectrum disorder with probiotic treatment are likely to be transitory [106]. Although a lot of effort has gone into the discovery of effective probiotics, there is surprisingly little evidence that these bacteria actually populate the intestine, at least in healthy adults [107]. It is probably a question of the right degree of immune–gut–brain stimulation because, as mentioned earlier, too much stimulation may lead to traveller's diarrhoea [76].

7.13. Genius

Having noted the changes in brain shape associated with potentially sub-optimal behaviour, the question remains: do these reflect active disease, or only characteristics? From the perspective of the early 21st century, it was considered that "dyslexia is a neurological disorder with a genetic origin", and that it "has lifelong persistence, reading retardation being merely one of its manifestations" [108]. Interestingly, if it is not a brain disease in itself, but rather a problem of differential brain development, it could be that obvious deficits such as dyslexia mask compensatory advantages elsewhere. Are dyslexics more "intelligent" (however this term is defined) than the average? Savant syndrome, the co-existence of both exceptional skill and neurological impairment, has been known about for a long time [109], as has the existence of an uneasy balance between psychopathology and creativity [110]. An entertaining tract highlighted by the title "The Mythconception of the Mad Genius" although replete with examples of actual "mad" "geniuses" (however those two terms are defined), still comes down in favour of madness being unnecessary for the expression of true genius [111]. In conclusion, although dyslexia, like hay fever, is a nuisance, it cannot readily be classed as a mere characteristic, even though it does not seem to lead on to more serious disease in its own right. Likewise, while it is hard to define genius as an actual mental illness, it is probably more "interesting" than a mere foible, and it still affords plenty of scope for further investigation.

7.14. Morning Sickness

While the dual inheritance hypothesis suggests that pregnancy involves a change in the rate of peristalsis mediated by the gut–brain axis, in reality we often see the appearance of mysterious so-called morning sickness—which can actually occur throughout the day [112]—as is the case for its rare but most debilitating version: hyperemesis gravidarum [113]. While an adaptive, prophylactic function for morning sickness has been proposed, and it is associated with lower rates of miscarriage, it is also known that there are some traditional societies in which morning sickness has never been observed [114]. However, perhaps not surprisingly for paternalistic societies, there does not seem to be any equivalent of Denis Burkitt's observations relating to his "Some Diseases Characteristic of Modern Western Civilization" [27] nor, indeed, of a female equivalent of John Bostock proclaiming the discovery of a new disease in front of her learned society [21]. In a sense, there is a connection between hay fever and the less serious forms of morning sickness in that, by and large, neither are life-threatening: a disturbance of normal function, rather than a disease. Indeed. it may be that pregnancy-associated sickness is related to microbiome-function deficiency, as an extreme case of the anxiety engendered by the changes in gut behaviour that the unconscious part of the brain is not wholly aware of. Of course, the consequences are similar to that perfectly reasonable, if dramatic, condition of traveller's diarrhoea [76], in which the sudden appearance of large numbers of unknown bacteria provokes a "get rid of them at all costs" response. Of course, the brain is not to know that these bacteria are probably harmless non-pathogens, just as a weak gut–brain axis may cause the relevant subconscious parts of the brain to mistake pregnancy for infection, with similarly unpleasant consequences.

7.15. Functional Gastrointestinal Disorders

As mentioned earlier, there is substantial overlap between coeliac disease and psychological manifestations [115], as there is with psychological problems and functional gastrointestinal disorders [4], perhaps indicating that both are aspects of microbiome-related deficiency disease, but with immune system input of variable seriousness. Consistent with the dual inheritance hypothesis, these problems are also found in the infant, a fact which carries its own special problems of diagnosis and treatment [116]. Significantly, it is also clear that mast cell-mediated inflammation is somehow involved, along with food allergy and bowel disorders such as Crohn's and ulcerative colitis [117], and such microbial-imbalance responses can also be seen in the small intestine [118]. Attempts to ameliorate irritable bowel syndrome by the introduction of bifidobacteria are not universally successful [119] and it is possible that peristalsis is improved by the placebo-like strengthening of the gut–brain axis after the interaction of these bacteria with the immune system of the host.

7.16. Straining

Following his ideas on low-fibre diets causing faecal arrest, among other complaints, Denis Burkitt blamed disorders like haemorrhoids, hiatus hernia, and varicose veins on intra-abdominal pressures rising above 200 mm Hg during straining on defecation [27]. An alternative explanation for such complaints may result from a lack of coordination following breakdown of the gut–brain axis. Defecation is a socially sensitive act and, as such, requires approval from higher management, i.e., the brain. The act itself is a reflex movement, requiring coordination of peristalsis and sphincter opening; however, at some point, permission to proceed, if granted by the brain, must be communicated to the relevant muscles through the gut–brain axis. If that message is not received at the critical moment, straining leads to eventual damage. Of course, fear stimulates bowel action, possibly by strengthening the gut–brain axis, and this may link in with the psychological problems associated with functional gastrointestinal disorders [1].

7.17. Summary

The full range of Burkitt-like non-communicable diseases are now beginning to make their presence felt, with the all too obvious inclusion of mental health conditions in recent generations. At this time, the most important thing is to confirm or deny the existence of key unicellular eukaryotes in populations not suffering from non-communicable disease, as it is to be expected that such entities would be rare or non-existent in afflicted peoples. Similarly, it will be necessary to identify relevant semiochemicals. In terms of the amelioration of existing disease, while probiotics do not seem to exert their action by supplementation of the gut microbiota, they may help to allay symptoms by stimulation of the immune–gut–brain triangle.

References

1. Drossman, D.A. Functional gastrointestinal disorders: History, pathophysiology, clinical features and Rome IV. *Gastroenterology* **2016**, *150*, 1262–1279. [CrossRef] [PubMed]
2. American Psychiatric Association. *Diagnostic and Statistical Manual of Mental Disorders*, 5th ed.; American Psychiatric Publishing: Arlington, VA, USA, 2013.
3. Strachan, D.P. Hay fever, hygiene and household size. *BMJ* **1989**, *299*, 1259–1260. [CrossRef] [PubMed]
4. Barker, D.J. The fetal and infant origins of adult disease. *BMJ* **1990**, *301*, 1111. [CrossRef] [PubMed]

5. Lenzi, J.; Messina, R.; Rosa, S.; Iommi, M.; Rucci, P.; Fantini, M.P.; Di Bartolo, P. A multi-state analysis of disease trajectories and mental health transitions in patients with type 2 diabetes: A population-based retrospective cohort study utilizing health administrative data. *Diabetes Res. Clin. Pract.* **2024**, *209*, 111561. [CrossRef] [PubMed]
6. Reilly, J.J.; El-Hamdouchi, A.; Diouf, A.; Monyeki, A.; Somda, S.A. Determining the world-wide prevalence of obesity. *Lancet* **2018**, *39*, 1773–1774. [CrossRef] [PubMed]
7. Sandercock, G.R.H.; Cohen, D.D. Temporal trends in muscular fitness of English 10-year-olds 1998–2014: An allometric approach. *J. Sci. Med. Sports* **2019**, *22*, 201–205. [CrossRef] [PubMed]
8. Elizabeth, L.; Machado, P.; Zinöcker, M.; Baker, P.; Lawrence, M. Ultra-processed foods and health outcomes: A narrative review. *Nutrients* **2020**, *12*, 1955. [CrossRef] [PubMed]
9. Resongles, E.; Dietze, V.; Green, D.C.; Harrison, R.M.; Ochoa-Gonzalez, R.; Tremper, A.H.; Weiss, D.J. Strong evidence for the continued contribution of lead deposited in the 20th century to the atmospheric environment of the London of today. *Proc. Natl. Acad. Sci. USA* **2021**, *118*, e21022791118. [CrossRef] [PubMed]
10. Needleman, H. The removal of lead from gasoline: Historical and personal reflections. *Environ. Res.* **2000**, *84*, 20–35. [CrossRef] [PubMed]
11. Shao, M.; Zhu, Y. Long term metal exposure changes gut microbiota of residents surrounding a mining and smelting area. *Sci. Rep.* **2020**, *10*, 4453. [CrossRef] [PubMed]
12. Protsiv, M.; Ley, C.; Lankester, J.; Hastie, T.; Parsonnet, J. Decreasing human body temperature in the United States since the Industrial Revolution. *eLife* **2020**, *9*, e49555. [CrossRef] [PubMed]
13. Lin, X.; Li, H. Obesity: Epidemiology, pathophysiology, and therapeutics. *Front. Endocrinol.* **2021**, *12*, 706978. [CrossRef] [PubMed]
14. Kuriakose, D.; Xiao, Z. Pathophysiology and treatment of stroke: Present status and future perspectives. *Int. J. Mol. Sci.* **2020**, *21*, 7609. [CrossRef] [PubMed]
15. Sun, H.; Warren, J.; Yip, J.; Ji, Y.; Hao, S.; Han, W.; Ding, Y. Factors affecting gallstone formation: A review of the literature. *Biomolecules* **2022**, *12*, 550. [CrossRef] [PubMed]
16. Pascart, T.; Lioté, F. Gout: State of the art after a decade of developments. *Rheumatology* **2019**, *58*, 27–44. [CrossRef] [PubMed]
17. Ferguson, L.D.; Molenberghs, G.; Verbeke, G.; Rahimi, K.; Rao, S.; McInnes, I.B.; McMurray, J.J.V.; Sattar, N.; Conrad, N. Gout and incidence of 12 cardiovascular diseases: A case-control study including 152 663 individuals with gout and 709 981 matched controls. *Lancet Rheumatol.* **2024**, *6*, E156–E167. [CrossRef] [PubMed]
18. Simon, K.; Hollander, G.A.; McMichael, A. Evolution of the immune system in humans from infancy to old age. *Proc. R. Soc. B* **2015**, *282*, 20143085. [CrossRef] [PubMed]
19. Yu, J.C.; Khodadadi, H.; Malik, A.; Davidson, B.; Salles, É.D.S.L.; Bhatia, J.; Hale, V.L.; Baban, B. Innate Immunity of Neonates and Infants. *Front. Immunol.* **2018**, *9*, 1759. [CrossRef] [PubMed]
20. Ahsan, H. Origins and history of autoimmunity-a brief review. *Rheumatol. Autoimmun.* **2022**, *3*, 9–14. [CrossRef]
21. Bostock, J. Case of a periodical affection of the eyes and chest. *Med. Chir. Trans.* **1819**, *10*, 161–165. [CrossRef] [PubMed]

22. Yang, L.; Fu, J.; Zhou, Y. Research progress in atopic march. *Front. Immunol.* **2020**, *11*, 1907. [CrossRef] [PubMed]
23. Peters, R.L.; Krawiec, M.; Koplin, J.J.; Santos, A.F. Update on food allergy. *Pediatr. Allergy Immunol.* **2021**, *32*, 647–657. [CrossRef] [PubMed]
24. Ugai, T.; Sasamoto, N.; Lee, H.Y.; Ando, M.; Song, M.; Tamimi, R.M.; Kawachi, I.; Campbell, P.T.; Giovannucci, E.L.; Weiderpass, E.; et al. Is early-onset cancer an emerging global epidemic? Current evidence and future implications. *Nat. Rev. Clin. Oncol.* **2022**, *19*, 656–673. [CrossRef] [PubMed]
25. Riquelme, E.; Zhang, Y.; Zhang, L.; Montiel, M.; Zoltan, M.; Dong, W.; Quesada, P.; Sahin, I.; Chandra, V.; Lucas, S.A.; et al. Tumor microbiome diversity and composition influence pancreatic cancer outcomes. *Cell* **2019**, *178*, 795–806. [CrossRef] [PubMed]
26. Barot, S.V.; Sangwan, N.; Nair, K.G.; Schmit, S.L.; Xiang, S.; Kamath, S.; Liska, D.; Khorana, A.A. Distinct intratumoral microbiome of young-onset and average-onset colorectal cancer. *eBioMedicine* **2024**, *100*, 104980. [CrossRef] [PubMed]
27. Burkitt, D.P. Some diseases characteristic of modern western civilization. *Br. Med. J.* **1973**, *1*, 274–278. [CrossRef] [PubMed]
28. Lindfors, K.; Ciacci, C.; Kurppa, K.; Lundin, K.E.A.; Makharia, G.K.; Mearin, M.L.; Murray, J.A.; Verdu, E.F.; Kaukinen, K. Coeliac disease. *Nat. Rev. Dis. Primers* **2019**, *5*, 3. [CrossRef] [PubMed]
29. Makharia, G.K.; Chauhan, A.; Singh, P.; Ahuja, V. Review article: Epidemiology of coeliac disease. *Aliment. Pharmacol. Ther.* **2022**, *56*, S3–S17. [CrossRef] [PubMed]
30. Ciccocioppo, R.; Kruzliak, P.; Cangemi, G.C.; Pohanka, M.; Betti, E.; Lauret, E.; Rodrigo, L. The spectrum of differences between childhood and adulthood celiac disease. *Nutrients* **2015**, *7*, 8733–8751. [CrossRef] [PubMed]
31. Reunala, T.; Salmi, T.T.; Hervonen, K.; Kaukinen, K.; Collin, P. Dermatitis herpetiformis: A common extraintestinal manifestation of coeliac disease. *Nutrients* **2018**, *10*, 602. [CrossRef] [PubMed]
32. Smyth, D.J.; Plagnol, V.; Walker, N.M.; Cooper, J.D.; Downes, K.; Yang, J.H.; Howson, J.M.; Stevens, H.; McManus, R.; Wijmenga, C.; et al. Shared and distinct genetic variants in type 1 diabetes and celiac disease. *N. Engl. J. Med.* **2008**, *359*, 2767–2777. [CrossRef] [PubMed]
33. Kylökäs, A.; Kaukinen, K.; Huhtala, H.; Collin, P.; Mäki, M.; Kurppa, K. Type 1 and type 2 diabetes in celiac disease: Prevalence and effect on clinical and histological presentation. *BMC Gastroenterol.* **2016**, *16*, 76. [CrossRef] [PubMed]
34. Bodkhe, R.; Shetty, S.A.; Dhotre, D.P.; Verma, A.K.; Bhatia, K.; Mishra, A.; Kaur, G.; Pande, P.; Bangarusamy, D.K.; Santosh, B.P.; et al. Comparison of small gut and whole gut microbiota of first-degree relatives with adult celiac disease patients and controls. *Front. Microbiol.* **2019**, *10*, 164. [CrossRef] [PubMed]
35. Leonard, M.M.; Valitutti, F.; Karathia, H.; Pujolassos, M.; Kenyon, V.; Fanelli, B.; Troisi, J.; Subramanian, P.; Camhi, S.; Colucci, A.; et al. Microbiome signatures of progression toward celiac disease onset in at-risk children in longitudinal prospective cohort study. *Proc. Natl. Acad. Sci. USA* **2021**, *118*, e2020322118. [CrossRef] [PubMed]
36. Alkhiari, R. Psychiatric and neurological manifestations of celiac disease in adults. *Cureus* **2023**, *15*, e35712. [CrossRef] [PubMed]

37. Pryce, C.R.; Fontana, A. Depression in autoimmune diseases. *Curr. Top. Behav. Neurosci.* **2017**, *31*, 139–154. [PubMed]
38. Lee, S.Y.; Pearce, E.N. Hyperthyroidism: A review. *JAMA* **2023**, *330*, 1472–1483. [CrossRef] [PubMed]
39. Di Matteo, A.; Bathon, J.M.; Emery, P. Rheumatoid arthritis. *Lancet Semin.* **2023**, *402*, P2019–P2033. [CrossRef] [PubMed]
40. Altieri, C.; Speranza, B.; Corbo, M.R.; Sinigaglia, M.; Bevilacqua, A. Gut-microbiota, and multiple sclerosis: Background, evidence, and perspectives. *Nutrients* **2023**, *15*, 942. [CrossRef] [PubMed]
41. Soldan, S.S.; Lieberman, P.M. Epstein-Barr virus and multiple sclerosis. *Nat. Rev. Microbiol.* **2023**, *21*, 51–64. [CrossRef] [PubMed]
42. Khan, S.A.; Das, P.R.; Nahar, Z.; Dewan, S.N.R. An updated review on Guillain-Barré syndrome: Challenges in infection prevention and control in low- and middle-income countries. *SAGE Open Med.* **2024**, *12*, 20503121241239538. [CrossRef] [PubMed]
43. Sista, S.B.; Crum, B.; Aboseif, A.; Devine, M.F.; Zekeridou, A.; Hammami, M.B.; Rezk, M.M.; Truffert, A.; Lalive, P.H.; Kunchok, A.; et al. Motor-neurone-disease-like phenotype associated with IgLON5 disease. *J. Neurol.* **2022**, *269*, 6139–6144. [CrossRef] [PubMed]
44. Callard, F.; Perego, E. How and why patients made Long COVID. *Soc. Sci. Med.* **2021**, *268*, 113426. [CrossRef] [PubMed]
45. Wong, T.L.; Weitzer, D.J. Long COVID and myalgic encephalomyelitis/chronic fatigue syndrome (ME/CFS)—A systematic review and comparison of clinical presentation and symptomatology. *Medicina* **2021**, *57*, 418. [CrossRef] [PubMed]
46. Shorter, E. Chronic fatigue in historical perspective. *Ciba Found. Symp.* **1993**, *173*, 6–16. [PubMed]
47. Sharma, C.; Bayry, J. High risk of autoimmune diseases after COVID-19. *Nat. Rev. Rheumatol.* **2023**, *19*, 399–400. [CrossRef] [PubMed]
48. Dotan, A.; Muller, S.; Kanduc, D.; David, P.; Halpert, G.; Shoenfeld, Y. The SARS-CoV-2 as an instrumental trigger of autoimmunity. *Autoimmun. Rev.* **2021**, *20*, 102792. [CrossRef] [PubMed]
49. Novak, P. Post COVID-19 syndrome associated with orthostatic cerebral hypoperfusion syndrome, small fibre neuropathy and benefit of immunotherapy: A case report. *eNeurologicalSci* **2020**, *21*, 100276. [CrossRef] [PubMed]
50. Molloy, E.J.; Nakra, N.; Gale, C.; Dimitriades, V.R.; Lakshminrusimha, S. Multisystem inflammatory syndrome in children (MIS-C) and neonates (MIS-N) associated with COVID-19: Optimizing definition and management. *Pediatr. Res.* **2023**, *93*, 1499–1508. [CrossRef] [PubMed]
51. Patel, P.; DeCuir, J.; Abrams, J.; Campbell, A.P.; Godfred-Cato, S.; Belay, E.D. Clinical characteristics of multisystem inflammatory syndrome in adults: A systematic review. *JAMA Netw. Open* **2021**, *4*, e2126456. [CrossRef] [PubMed]
52. Cervia-Hassler, C.; Brüningk, S.C.; Hoch, T.; Fan, B.; Muzio, G.; Thompson, R.C.; Ceglarek, L.; Meledin, R.; Westermann, P.; Emmenegger, M.; et al. Persistent complement dysregulation with signs of thromboinflammation in active Long COVID. *Science* **2024**, *383*, eadg7942. [CrossRef] [PubMed]

53. Filler, K.; Lyon, D.; Bennet, J.; McCain, N.; Elswick, R.; Lukkahatai, N.; Saligan, L.N. Association between mitochondrial dysfunction and fatigue: A review of the literature. *BBA Clin.* **2014**, *1*, 12–23. [CrossRef] [PubMed]
54. Al-Aly, Z.; Topol, E. Solving the puzzle of Long Covid. *Science* **2024**, *383*, 830–832. [CrossRef] [PubMed]
55. Appelman, B.; Charlton, B.T.; Goulding, R.P.; Kerkhoff, T.J.; Breedveld, E.A.; Noort, W.; Offringa, C.; Bloemers, F.W.; van Weeghel, M.; Shomakers, B.V.; et al. Muscle abnormalities worsen after post-exertional malaise in long COVID. *Nat. Commun.* **2024**, *15*, 17. [CrossRef] [PubMed]
56. Fuxench, Z.C.C.; Wan, J.; Wang, S.; Syed, M.N.; Shin, D.B.; Abuabara, K.; Gelfand, J.M. Risk of inflammatory bowel disease in patients with atopic dermatitis. *JAMA Dermatol.* **2023**, *159*, 1085–1092. [CrossRef] [PubMed]
57. Ocampo, D.V.; Gladman, D. Psoriatic arthritis. *F1000Research* **2019**, *8*, F1000 Faculty Rev-1665. [CrossRef] [PubMed]
58. Segre, J.A. What does it take to satisfy Koch's postulates two centuries later? Microbial genomics and Propionibacteria acnes. *J. Investig. Dermatol.* **2013**, *133*, 2141–2142. [CrossRef] [PubMed]
59. Bowe, W.P.; Logan, A.C. Acne vulgaris, probiotics and the gut-brain-skin axis—Back to the future? *Gut Pathog.* **2011**, *3*, 1. [CrossRef] [PubMed]
60. Ríos-Covian, D.; Langella, P.; Martín, R. From short- to long-term effects of C-section delivery on microbiome establishment and host health. *Microorganisms* **2021**, *9*, 2122. [CrossRef]
61. Li, Z.; Liang, H.; Hu, Y.; Lu, L.; Zheng, C.; Fan, Y.; Wu, B.; Zou, T.; Luo, X.; Zhang, X.; et al. Gut bacterial profiles in Parkinson's disease: A systematic review. *CNS Neurosci. Ther.* **2023**, *29*, 140–157. [CrossRef] [PubMed]
62. Seo, D.-O.; Holtzman, D.M. Current understanding of the Alzheimer's disease-associated microbiome and therapeutic strategies. *Exp. Mol. Med.* **2024**, *56*, 86–94. [CrossRef] [PubMed]
63. Chen, Y.; Xu, J.; Chen, Y. Regulation of neurotransmitters by the gut microbiota and effects on cognition in neurological disorders. *Nutrients* **2021**, *13*, 2099. [CrossRef] [PubMed]
64. Aggarwal, M.; Alkhayyat, M.; Abou Saleh, M.; Sarmini, M.T.; Singh, A.; Garg, R.; Garg, P.; Mansoor, E.; Padival, R.; Cohen, B.L. Alzheimer disease occurs more frequently in patients with inflammatory bowel disease: Insight from a nationwide study. *J. Clin. Gastroenterol.* **2023**, *57*, 501–507. [CrossRef] [PubMed]
65. Lee, C.H.; Han, K.; Lee, H.J.; Yu, H.; Kim, S.; Choi, K.; Koh, S.-J.; Im, J.P.; Kim, J.S. Migraine is associated with the development of adult patients with inflammatory bowel disease: A nationwide, population-based study. *Sci. Rep.* **2024**, *14*, 1157. [CrossRef] [PubMed]
66. Joham, A.E.; Norman, R.J.; Stener-Victoria, E.; Legro, R.S.; Franks, S.; Moran, L.J.; Boyle, J.; Teede, H.J. Polycystic Ovary Syndrome. *Lancet Diabetes Endocrinol.* **2022**, *10*, 668–680. [CrossRef] [PubMed]

67. Liu, J.; Teng, Z.; Xie, H.; Yuan, H.; Liu, M.; Chen, J.; Tang, H.; Xiong, H.; Wu, H.; Huang, J. Prevalence and characteristics of polycystic ovarian syndrome in patients with bipolar disorder. *J. Affect. Disord.* **2023**, *340*, 387–395. [CrossRef] [PubMed]
68. Sadeghi, M.H.; Adelhi, I.; Calina, D.; Docea, A.O.; Mousavi, T.; Daniali, M.; Nikfar, S.; Tsatsakis, A.; Abdollahi, M. Polycystic ovary syndrome: A comprehensive review of pathogenesis, management, and drug repurposing. *Int. J. Mol. Sci.* **2022**, *23*, 583. [CrossRef] [PubMed]
69. Silano, M.; Vincentini, O.; De Vincenzi, M. Toxic, immunostimulatory and antagonist gluten peptides in celiac disease. *Curr. Med. Chem.* **2009**, *16*, 1489–1498. [CrossRef] [PubMed]
70. Gasbarrini, G.; Rickards, O.; Martinez-Labarga, C.; Pacciani, E.; Chilleri, F.; Laterza, L.; Marangi, G.; Scaldaferri, F.; Gasbarrini, A. Origin of celiac disease: How old are predisposing haplotypes? *World J. Gastroenterol.* **2012**, *18*, 5300–5304. [CrossRef]
71. Kleinjans, M.; Schneider, C.V.; Bruns, T.; Strnad, P. Phenome of coeliac disease vs. inflammatory bowel disease. *Sci. Rep.* **2022**, *12*, 14572. [CrossRef] [PubMed]
72. Ashok, T.; Patni, N.; Fatima, M.; Lamis, A.; Siddiqui, S.W. Celiac disease and autoimmune thyroid disease: The two peas in a pod. *Cureus* **2022**, *14*, e26243. [CrossRef] [PubMed]
73. Balint, E.; Feng, E.; Giles, E.C.; Ritchie, T.M.; Qian, A.S.; Vahedi, F.; Montemarano, A.; Portillo, A.L.; Montiero, J.K.; Trigatti, B.L.; et al. Bystander activated $CD8^+$ T cells mediate neuropathology during viral infection via antigen-independent cytotoxicity. *Nat. Commun.* **2024**, *15*, 896. [CrossRef] [PubMed]
74. Chen, W.G.; Schloesser, D.; Arensdorf, A.M.; Simmons, J.M.; Cui, C.; Valentino, R.; Gnadt, J.W.; Nielsen, L.; St. Hilaire-Clarke, C.; Spraunce, V.; et al. The emerging science of interoception: Sensing, integrating, interpreting, and regulating signals within the self. *Trends Neurosci.* **2021**, *44*, 3–16. [CrossRef] [PubMed]
75. Glans, M.R.; Thelin, N.; Humble, M.B.; Elwin, M.; Bejerot, S. The relationship between generalised joint hypermobility and autism spectrum disorder in adults: A large, cross-sectional, case control comparison. *Front. Psychiatry* **2022**, *12*, 803334. [CrossRef] [PubMed]
76. Giddings, S.L.; Stevens, A.M.; Leung, D.T. Traveler's diarrhea. *Med. Clin. N. Am.* **2016**, *100*, 317–330. [CrossRef] [PubMed]
77. Szuhany, K.L.; Simon, N.M. Anxiety disorders: A review. *JAMA* **2022**, *328*, 2431–2445. [CrossRef] [PubMed]
78. Nimmons, D.; Bazo-Alvarez, J.C.; Avgerinou, C.; Hayes, J.; Osborn, D.; Cooper, C.; Petersen, I.; Walters, K. Exploring the co-occurrence of depression, anxiety and insomnia symptoms, diagnoses and treatments in clinical care: Observational study using UK primary care data. *BJPsych Open* **2024**, *10*, e76. [CrossRef] [PubMed]
79. Xiang, H.; Xue, Y.; Wang, J.; Weng, Y.; Rong, F.; Peng, Y.; Ji, K. Cardiovascular alterations and management of patients with white coat hypertension: A meta-analysis. *Front. Pharmacol.* **2020**, *11*, 570101. [CrossRef] [PubMed]
80. Puckett, L.; Grayeb, D.; Khatri, V.; Cass, K.; Mehler, P. A comprehensive review of complications and new findings associated with anorexia nervosa. *J. Clin. Med.* **2021**, *10*, 2555. [CrossRef] [PubMed]

81. Tanne, J.H. Epidemic of loneliness threatens public health, says US surgeon general. *BMJ* **2023**, *381*, p1017. [CrossRef] [PubMed]
82. Blundell, E.; De Stavola, B.L.; Kellock, M.D.; Kelly, Y.; Lewis, G.; McMunn, A.; Nicholls, D.; Patalay, P.; Solmi, F. Longitudinal pathways between childhood BMI, body dissatisfaction, and adolescent depression: An observational study using the UK Millennium Cohort Study. *Lancet Psychiatry* **2024**, *11*, 47–55. [CrossRef] [PubMed]
83. López-Gil, J.F.; García-Hermoso, A.; Smith, L.; Firth, J.; Trott, M.; Mesas, A.E.; Jiménez-López, E.; Gutiérrez-Espinoza, H.; Tárraga-López, P.J.; Victoria-Montesinos, D. Global proportion of disordered eating in children and adolescents: A systematic review and meta-analysis. *JAMA Pediatr.* **2023**, *177*, 363–372. [CrossRef] [PubMed]
84. Cooper, K.; Russell, A.; Mandy, W.; Butler, C. The phenomenology of gender dysphoria in adults: A systemic review and meta-synthesis. *Clin. Psychol. Rev.* **2020**, *80*, 101875. [CrossRef] [PubMed]
85. Block, J. Gender dysphoria in young people is rising–and so is professional disagreement. *BMJ* **2023**, *380*, 382. [CrossRef] [PubMed]
86. Jassi, A.; Krebs, G. Body dysmorphic disorder: Reflections on the last 25 years. *Clin. Child. Psychol. Psychiatry* **2021**, *26*, 3–7. [CrossRef] [PubMed]
87. Yan, Z.; Lei, Y.; Zhao, P.; Zhang, D.; Shen, J.; Zhang, G.; Wei, R.; Liu, H.; Liu, X.; He, Y.; et al. Natural mating ability is associated with gut microbiota composition and function in captive male giant pandas. *Ecol. Evol.* **2024**, *14*, e11189. [CrossRef] [PubMed]
88. Reese, A.T.; Chadaideh, K.S.; Diggins, C.E.; Schell, L.D.; Beckel, M.; Callahan, P.; Ryan, R.; Thompson, M.E.; Carmody, R.N. Effects of domestication on the gut microbiota parallel those of human industrialization. *eLife* **2021**, *10*, e60197. [CrossRef] [PubMed]
89. Choy, O.; Raine, A.; Schug, R. Larger striatal volume is associated with increased adult psychopathy. *J. Psychiatr. Res.* **2022**, *149*, 185–193. [CrossRef] [PubMed]
90. Postema, M.C.; van Rooij, D.; Anagnostou, E.; Arango, C.; Auzias, G.; Behrmann, M.; Filho, G.B.; Calderoni, S.; Calvo, R.; Daly, E.; et al. Altered structural brain asymmetry in autism spectrum disorder in a study of 54 datasets. *Nat. Communcations* **2019**, *10*, 4958. [CrossRef] [PubMed]
91. Lynch, C.J.; Elbau, I.G.; Ng, T.; Avaz, A.; Zhu, S.; Wolk, D.; Manfredi, N.; Johnson, M.; Chang, M.; Chou, J.; et al. Frontostriatal salience network expansion in individuals with depression. *Nature*, **2024**, *633*, 624–633. [CrossRef] [PubMed]
92. Williams, J.A.; Burgess, S.; Suckling, J.; Lalousis, P.A.; Batool, F.; Griffiths, S.L.; Palmer, E.; Karwath, A.; Barsky, A.; Gkoutos, G.V.; et al. Inflammation and brain structure in schizophrenia and other neuropsychiatric disorders: A Mendelian randomisation study. *JAMA Psychiatry* **2022**, *79*, 498–507. [CrossRef] [PubMed]
93. Garcia-Gutierrez, E.; Narbad, A.; Rodriguez, J.M. Autism spectrum disorder associated with gut microbiota at immune, metabolomic, and neuroactive level. *Front. Neurosci.* **2020**, *14*, 578666. [CrossRef] [PubMed]
94. Budu-Aggrey, A.; Joyce, S.; Davies, N.M.; Paternoster, L.; Munafò, M.R.; Brown, S.J.; Evans, J.; Sallis, H.M. Investigating the causal relationship between allergic disease and mental health. *Clin. Exp. Allergy* **2021**, *51*, 1449–1458. [CrossRef] [PubMed]
95. Chiarotti, F.; Venerosi, A. Epidemiology of autism spectrum disorders: A review of worldwide prevalence since 2014. *Brain Sci.* **2020**, *10*, 274. [CrossRef] [PubMed]

96. Sandin, S.; Lichtenstein, P.; Kuja-Halkola, R.; Hultman, C.; Larsson, H.; Reichenberg, A. The heritability of autism spectrum disorder. *JAMA* **2017**, *318*, 1182–1184. [CrossRef] [PubMed]
97. Bundgaard-Nielson, C.; Knudson, J.; Leutscher, P.D.C.; Lauritsen, M.B.; Nyegaard, M.; Hagstrøm, S.; Sørensen, S. Gut microbiota profiles of autism spectrum disorder and attention deficit/hyperactivity disorder: A systematic literature review. *Gut Microbes* **2020**, *11*, 1172–1187. [CrossRef] [PubMed]
98. Zhou, J.; Wu, X.; Xiang, T.; Liu, F.; Gao, H.; Tong, L.; Yan, B.; Li, Z.; Zhang, C.; Wang, L.; et al. Dynamical alterations of brain function and gut microbiome in weight loss. *Front. Cell. Infect. Microbiol.* **2023**, *13*, 1269548. [CrossRef] [PubMed]
99. DeCarli, C.; Maillard, P.; Pase, M.P.; Beiser, A.S.; Kojis, D.; Satizabal, C.L.; Himali, J.J.; Aparicio, H.J.; Fletcher, E.; Seshadri, S. Trends in intracranial and cerebral volumes of Framingham heart study participants born 1930 to 1970. *JAMA Neurol.* **2024**, *81*, 471–480. [CrossRef] [PubMed]
100. Grajek, M.; Krupa-Kotara, K.; Białek-Dratwa, A.; Sobczyk, K.; Grot, M.; Kowalski, O.; Staśkiewicz, W. Nutrition and mental health: A review of current knowledge about the impact of diet on mental health. *Front. Nutr.* **2022**, *9*, 943998. [CrossRef] [PubMed]
101. de Craen, A.J.; Kaptchuck, T.J.; Tijssen, J.G.; Kleijnen, J. Placebos and placebo effects in medicine: Historical overview. *J. R. Soc. Med.* **1999**, *92*, 511–515. [CrossRef] [PubMed]
102. Barsky, A.J.; Saintfort, R.; Rogers, M.P.; Borus, J.F. Nonspecific medication side effects and the nocebo phenomenon. *JAMA* **2002**, *87*, 622–627. [CrossRef] [PubMed]
103. Cohen, S.I. Psychosomatic death: Voodoo death in a modern perspective. *Integr. Psychiatry* **1985**, *3*, 46–51.
104. Hrobjartsson, A.; Kaptchuk, T.J.; Miller, F.G. Placebo effect studies are susceptible to response bias and other types of biases. *J. Clin. Epidemiol.* **2011**, *64*, 1223–1229. [CrossRef] [PubMed]
105. Charlesworth, J.E.; Petkovic, G.; Kelley, J.M.; Hunter, M.; Onakpoya, I.; Roberts, N.; Miller, F.G.; Howick, J. Effects of placebos without deception compared with no treatment: A systematic review and meta-analysis. *J. Evid.-Based Med.* **2017**, *10*, 97–107. [CrossRef] [PubMed]
106. Martínez-González, A.E.; Andreo-Martínez, P. Prebiotics, probiotics and fecal microbiota transplantation in autism: A systematic review. *Rev. Psiquiatr. Salud Ment. (Engl. Ed.)* **2020**, *13*, 150–164. [CrossRef] [PubMed]
107. Khalesi, Z.; Bellisimo, N.; Vandelanotte, C.; Williams, S.; Stanley, D.; Irwin, C. A review of probiotic supplementation in healthy adults: Helpful or hype? *Eur. J. Clin. Nutr.* **2019**, *73*, 24–37. [CrossRef] [PubMed]
108. Ramus, F.; Rosen, S.; Dakin, S.C.; Day, B.L.; Castellote, J.M.; White, S.; Frith, U. Theories of developmental disorder: Insights from a multiple case study of dyslexic adults. *Brain* **2023**, *126*, 841–865. [CrossRef] [PubMed]
109. Miller, L.K. The savant syndrome: Intellectual impairment and exceptional skill. *Psychol. Bull.* **1999**, *125*, 31–46. [CrossRef] [PubMed]
110. Thys, E.; Sabbe, B.; De Hert, M. Creativity and psychopathology: A systematic review. *Psychopathology* **2014**, *47*, 141–147. [CrossRef] [PubMed]

111. Dietrich, A. The mythconception of the mad genius. *Front. Psychol.* **2014**, *5*, 79. [CrossRef] [PubMed]
112. Gadsby, R.; Ivanova, D.; Trevelyan, E.; Hutton, J.L.; Johnson, S. Nausea and vomiting in pregnancy is not just 'morning sickness': Data from a prospective cohort study in the UK. *Br. J. Gen. Pract.* **2020**, *70*, e534–e539. [CrossRef] [PubMed]
113. Vinnars, M.-T.; Forslund, M.; Claesson, I.-M.; Hedman, A.; Peira, N.; Olofsson, H.; Wernersson, E.; Ulfsdottir, H. Treatments for hyperemesis gravidarum: A systematic review. *Acta Obstet. Gynecol. Scand.* **2023**, *103*, 13–29. [CrossRef] [PubMed]
114. Flaxman, S.M.; Sherman, P.W. Morning sickness: A mechanism for protecting mother and embryo. *Q. Rev. Biol.* **2000**, *75*, 113–148. [CrossRef] [PubMed]
115. Zingone, F.; Swift, G.L.; Card, T.R.; Sanders, D.S.; Ludvigsson, J.F.; Bai, J.C. Psychological morbidity of celiac disease: A review of the literature. *United Eur. Gastroenterol. J.* **2015**, *3*, 136–145. [CrossRef] [PubMed]
116. Hyams, J.S.; Di Lorenzo, C.; Saps, M.; Shulman, R.J.; Staiano, A.; van Tilburg, M. Childhood functional gastrointestinal disorders: Child/adolescent. *Gastroenterology* **2016**, *150*, 1456–1468. [CrossRef] [PubMed]
117. Wouters, M.M.; Vicario, M.; Santos, J. The role of mast cells in functional GI disorders. *Gut* **2016**, *65*, 155–168. [CrossRef] [PubMed]
118. Burns, G.L.; Talley, N.J.; Keely, S. Immune responses in the irritable bowel syndromes: Time to consider the small intestine. *BMC Med.* **2022**, *20*, 115. [CrossRef] [PubMed]
119. Pratt, C.; Campbell, M.D. The effect of bifidobacterium on reducing symptomatic pain in patients with irritable bowel syndrome: A systematic review. *Probiotics Antimicrob Proteins* **2020**, *12*, 834–839. [CrossRef] [PubMed]

8. Attempts at Amelioration

8.1. Disentangling Disease

While the "cutting edge" concepts of disease are moving towards microbial involvement, practical steps remain firmly grounded in 20th century validated methods. Of course, this leads to a tension at the new/old interface when, for example, probiotics are recommended for use alongside antibiotics [1]. In essence, oral antibiotics not only generate resistance in their bacterial targets, but they are also damaging to bystander microbes and incompatible with the idea of a diverse, fully functioning microbiome [2]. By contrast, phage therapy seems to be a valid microbiome-preserving technology, in principle at least. Equally, the probiotic concept, that specific added bacteria can improve the health of both the body and mind, perhaps falls down on the fact mentioned earlier, that people seem to be otherwise healthy after undergoing colectomy [3] and are even capable of bearing their own children with minimal disturbance to their reproductive capacity [4]. In fact, the best way to explain the apparent lack of serious microbiome-related problems after colectomy is to say that these patients are actually no worse than the bulk of people who already have a malfunctioning microbiome. Accordingly, it is the purpose of this chapter to re-interpret current treatment methods in the light of this hypothesis, while the next and last section, Chapter 9, perhaps offers a way to ensure that our as-yet-unborn children's children will eventually be free from these characteristic non-communicable diseases.

8.2. Antibiotic Resistance

Not surprisingly, the bulk of current antibiotic discovery effort is designed to counter the rising threat of antibiotic resistance [5] but there have been some efforts to minimise the problem of antibiotics reducing microbiome diversity [6]. However, it seems clear that oral antibiotics will cause problems if taken either during childhood or in pregnancy, if only by reducing their ability to express mobile genetic elements. In other circumstances, however, antibiotics may not be the major reason for the overall loss of microbial diversity. As normally prescribed, antibiotics work alongside the immune system to resolve infections, but the immune system is less effective inside the gut lumen. Equally, while the lymphatic system clears away debris from the site of infection, the lack of such clearance from the gut lumen allows the reassembly of microbes according to the viral shunt mechanism [7]. Accordingly, it is possible that antibiotics simply contribute to the normal phage-induced turnover within the gut lumen. Furthermore, it is considered possible that key microbes held within the appendix may act as a reservoir to repopulate the intestinal microbiome sometime after the course has finished [8]. According to the dual inheritance hypothesis, the most dangerous time for antibiotic use will be around pregnancy and childbirth, when key microbes must be physically transferred from mother to child. Even this caveat will not apply if the antibiotics are delivered by injection, in which case the effect on the gut, and hence maternal microbial transfer, is likely to be minimal. Of course, injected antibiotics, where the bulk of the dose does not interfere with the normal gut microbiota, are less likely to induce antibiotic resistance in the first place.

8.3. Phage Therapy

Increasing levels of the genes responsible for antibiotic resistance have led to the resurrection of an old idea: that pathogen-specific phage viruses could be generated and injected back into infected individuals [9]. Paradoxically, however, a lot of the work on alternative antimicrobial agents, including phage technology, has been performed on farmed animals [10], part of the rationale being that husbandry is responsible for significant levels of antibiotic resistance, but also because much safety work has to be carried out in animals before it can be used in humans. Interestingly, it has long been known that treatment of farmed animals with antibiotics increases their rate of growth [11], presumably by interfering with microbiome function, as well as causing immune system problems in pet animals [12]. As increased meat production is clearly a better idea than microbiome-related disease in people and our pets, the ultimate aim must be the opposite: to retain antibiotics for meat production while using the alternatives for our pets and ourselves.

8.4. Probiotics and Their Kin

As the medical/scientific world began to realise that the range of non-communicable diseases may be connected to the low diversity of microbes in the large intestine [13], the logical—some may say naïve—response was to simply add more, in the shape of bacteria-enhanced "probiotic" foodstuffs. Of course, this approach has a long history, perhaps starting with the observation of the increased lifespan of yoghurt-eating Bulgarian farmers, a point taken up by the Nobel prize-winning Élie Metchnikoff in the early 20th century and his studies on *Lactobacillus* consumption [14]. After the events of the middle 20th century and the later rise in non-communicable disease, these ideas have been revised and re-introduced in a variety of ways, such as probiotics (meaning "for life", i.e., the bacterial entities themselves), prebiotics (substances that are known to feed the microbes, normally dietary fibre), and synbiotics (a combination of pro- and prebiotics), all of which need to be precisely defined [15]. In a similar vein, the apparent ability of added bacteria to improve mental health led to the appearance of the word "psychobiotics" [16]. Current emphasis seems to be shifting to the compounds that these probiotics actually produce, the "postbiotics," in the hope that some of these will be the actual medicinal agents [17]. The final word in our "biotics" collection is "paraprobiotics," a term used for probiotics whose active bacterial agents have been deliberately killed during the manufacturing process, thus rendering a more stable final product from the quality control point of view [18]. In principle these no-longer-living molecular entities are safer than probiotics themselves, as they contain no viable bacteria; however, they still engage with the immune system and thereby, hopefully, express their benefits. Interestingly, in these latter two terms there is an echo of the early 20th century "transforming factor" observations of Frederick Griffith, referred to earlier, in that degraded bacteria can still pass on their characteristic features by what later became known as horizontal gene transfer [19].

Accordingly, in this way paraprobiotics could, indeed, pass on their value to the customer. Sadly, after all this "biotics" work, the conclusion is that "more research is needed," at least from a sports recovery perspective [20].

8.5. Fermentation

While it has been known since the earliest times [21], food fermentation has undergone a renaissance over recent decades [22], down to carefully looking at the precise detail of polysaccharide chemistry [23], and at the microbes themselves [24]. Of course, it is the job of the medical scientists to sort out health claims from health benefits, although controlled human studies on fermented foods themselves can be hard to arrange [25].

8.6. The Promise of Probiotics

In essence, foodstuffs containing adequate levels of suitably antigenic materials, be they harmless live bacteria or fragments thereof, will initially interact with the immune system. In turn, this system will send signals to both brain and gut, strengthening the gut–brain axis and thereby, perhaps, temporarily alleviating some non-communicable conditions. While the placebo effect itself is normally considered to be stimulated solely by talk [26], the ability of "biotics" to simultaneously engage with brain, gut, and immune system, increases their potency and, presumably, helps account for the popularity of probiotics with the public. Needless to say, there will be many factors involved with this subtle "treatment," perhaps requiring fine tuning between patient and medicine, with the threat that over-reactivity with live bacteria may strongly resemble the consequences of traveller's diarrhoea [27], an outcome that will not be good for sales!

8.7. "Blind" Medication

Beyond probiotics and functional foods in general, it is probably fair to say that the medical/scientific world has been wrong-footed by the sudden rise in non-communicable disease. As an example, even the current frontrunner against obesity, the glucagon-like peptide agonists [28], were discovered "blind", i.e., without a full understanding of the reasons out-of-control weight gain occurs in the first place [29]. Similarly, while metformin has a long and distinguished career behind it [30], we are now stumbling across new uses to offset the effects of microbiome-function deficiency disease "beyond diabetes" [31]. Other chemicals are beginning to show promise for what can, perhaps, be best described as "lifestyle replacement therapies": exercise mimetics [32] and calorie reduction mimetics [33]. Only time and clinical trials will tell if these agents offer more than just profits for shareholders but, even if successful in reducing levels of non-communicable disease, it is important to remember that they only put us back to where we were before the advent of industrialisation. Essentially the same story applies to aspirin, from a Wonder of the Ancient World (as salicylic acid) to an effective antithrombotic in the modern [34]. Indeed, such is our lack of knowledge that it is not clear if aspirin is a medicinal agent in its own right, or if it is actually a prodrug for salicylic acid [35]. Nevertheless, a new indication for aspirin is gaining greater interest in that, while its anti-metastatic properties have been established for number of years, if not decades, only now are medics beginning to overcome their scruples and tentatively suggesting its use alongside cancer therapy [36]. Sadly, as there are no actual cures for microbiome-function deficiency disease currently in sight, the emphasis must be on prevention in future generations or, at least, on preventing the current situation from getting any worse.

8.8. *The Effects of Exercise*

Of course, the economic world runs on the flow of capital so, when politicians belatedly realised that obesity and type 2 diabetes were placing an intolerable strain on health systems around the world, the answer was quite clear: encourage people to spend more money on "burning the fat off" by joining a gym. As we saw earlier, however, although this approach has been discussed as the "energy balance model" by scientists [37], there seems to be strong evidence for so-called compensation mechanisms being mobilised to defend against weight loss [38]. Although exercise does not easily reduce excess adiposity [39], it can help to both generate fitness, and to maintain the ideal weight and shape [40]. In addition, a direct comparison for the relief of depression and anxiety disorders by either medication or "running therapy" showed the latter to be substantially better for both physical and mental health [41]. Overall, it seems that a set of hormones called exerkines may be responsible for the majority of their benefits [42]. Finally, while the idea that exercise reduces cancer is hard to prove [43], the intriguing observation that GLP-1 receptor agonists are not only active against type 2 diabetes but also seem to reduce the incidence of colorectal cancer [44] suggests a mechanism: that the fasting-like reduction of circulating glucose may slow cancer growth simply by increasing competition for limited nutritional resources.

8.9. *Standard Human and Standard Microbiome*

Sadly, it seems that the slowing of peristalsis consequent upon microbiome-function deficiency disease is increasing in extent and severity with each new generation, producing what could called a "drift" of body statistics. To re-emphasise the findings of Chapter 3, this can most clearly be seen by the well-documented fact that, while people have been eating less in recent decades, they have also been increasing in weight [45] and decreasing in average body temperature, at least in the United States, since the Industrial Revolution [46]. The effect on children is particularly pertinent, with measurements of height, circumference, weight, and strength being carried out across different groups of ten-year-old children in 1998, 2008, and 2014. It was found that all dimensions expanded in equal ratios, including weight, leaving the body mass index unaltered but, worryingly, with decreasing strength [47]. Accordingly, the question of the correct statistics to represent a healthy human continues to perplex the medical profession, although the answer will lie within those populations not suffering from non-communicable disease, such as the microbially diverse Hadza, for example [48]. However, it is necessary to bear in mind that the "standard" ranges of weight, shape and, indeed, microbiome, will depend on local population variants, such as those adapted for polar or for tropical regions, for example. Indeed, it is important to note that, for a given population to be classed as a locally standard example of healthy *H. sapiens*, no member of that population should be found to be suffering from one of the Burkitt-like non-communicable diseases [49]. As an example, while older members of the so-called "Blue Zones" may represent standards for their local population body type [50], younger members may well be at risk of body shape drift due to pollution having caused subsequent microbiome-function deficiency disease.

8.10. Shape Control

It seems reasonable to suppose that people who seem to be healthy can reduce their chances of developing some of these microbiome-related degenerative disorders if they can maintain their weight and body shape within whatever their optimum range happens to be. As related in Chapter 3, these questions came to the fore in Europe in the 19th century, alongside rising levels of obesity and, indeed, hay fever [51]. Ultimately the most famous answer was that by Adolphe Quetelet, later known as body mass index, or BMI [52]. At the time it was known that this mathematical weight/surface area ratio is not really suitable for individual use, as too wide a range of anywhere between 18.5 and 24.9 kg/m^2 could be defined as healthy. The problem is that weight on its own cannot distinguish between healthy muscle and unhealthy visceral fat, for example, so other factors must be taken into account. Sadly, simpler measures, such as waist circumference, or the ratios waist-to-hip or waist-to-height, are reported to be either no better or of unproven value, and were only determined within Caucasian populations [53]. Meanwhile, a composite measure is under consideration specifically for individual use: "lipid accumulation product" is a comparison of a simple waist circumference measurement combined with blood levels of triglyceride, acting as a marker for visceral fat and the likelihood of type 2 diabetes [54]. While the addition of a blood test makes this too expensive for general screening, a more expensive method that is only suitable for research involves DEXA (Dual-Energy X-ray Absorptivity), which measures the absorption of X-radiation at two energy levels [55]. This method has been applied to a large group of younger individuals, following their growth from 9 to 24 years of age and comparing body fat distribution with various physical measures, including BMI. The best predictor of good health turned out to be the ratio of waist circumference to height, abbreviated as WHtR, in which about 0.5 corresponded to seemingly good levels of body fat compared to muscle, essentially independently of age or gender. In turn, this indicates that WHtR is probably an inexpensive and superior replacement for BMI over a wide age-range [56]. Reinforcing these observations, the National Institute for Health and Care Excellence (NICE) of the United Kingdom has recently issued a draft guidance encouraging people to maintain a waist diameter of less than half their height [57]. As seen in a previous paragraph, although exercise can help to maintain shape, it seems that some kind of diet control is needed to remove any excess fat storage before exercise can begin to do its job.

8.11. *The Perils of Dietary Intervention*

As described by Figure 8 (Chapter 6), the key point about the dual inheritance hypothesis is that a microbiome, fully functioning from birth, is capable of accelerating gut peristalsis to supply itself with nutrition, only being over-ruled when its host needs the food to overcome adverse conditions such as famine. Conversely, the failure to establish gut–brain connections in the early years forfeits this ability, increasing intestinal transit time and leaving the host vulnerable to disorders resulting from overabsorption of nutrition. At this point it must be remembered that complicating factors such as coeliac disease may disrupt this absorption and give the superficial appearance of health. Nevertheless, while mental health problems may be invisible, even to the sufferer, and immune system malfunction may be controlled by antihistamines or emollient cream preparation, for example, food-related gastrointestinal disorders are uncomfortable and/or embarrassing, and obesity is visible to all. Accordingly, most non-pharmaceutical effort is focused on weight loss interventions, but without any universally accepted underlying theory as to the reasons the problem occurs in the first place. Diet must always be used with care, however, as there exists a sizeable number of people who might be pushed into disordered eating, an increasing problem in the young [58], or even outright anorexia nervosa, an all too common condition which is more dangerous to the individual than type 2 diabetes [59]. Sadly, the zeitgeist of the modern world is exactly the opposite of what is needed, as we are invariably encouraged to eat more food containing more easily digested substances.

8.12. *The Ultraprocessed "Diet"*

A large proportion of the food eaten nowadays, at least in the richer countries of the world, has been "ultraprocessed": disassembled and reconstituted, often with added sweeteners and emulsifiers. As an increasing proportion of people in the richer world are overweight and obese, naturally, this level of processing has come under scrutiny [60]. Unfortunately, it is difficult to find the exact reasons why the level of processing should cause weight gain, making "ultra-processing" seem to be an artificial concept, invented merely for want of anything better. Sadly, an attempt to avoid such foods by providing "healthy groceries" sufficient for ten meals per week, per household, to type 2 diabetic food-insecure people, did not improve their level of glycaemic control [61]. Rather than the intrinsic nature of over-processed foods being to blame, perhaps the best way to describe the unhealthy nature of such foods is also the simplest: that they encourage overeating. In one experiment, 20 weight-stable adults were randomly assigned into two groups, eating similar ad lib diets of either "normal" or "ultraprocessed" food (however defined) for 2 weeks, then swapping for a further 2 weeks, but always under close supervision. The answer was clear-cut, as both groups ate more food and gained more weight during the ultraprocessed leg of the experiment, indicating hyper-palatability as the cause of the weight gain [62]. At this point it is important to emphasise that overeating is only a problem when the population suffers from microbiome-function deficiency disease as, otherwise, excess nutrition would contribute to microbial overgrowth and, eventually, be excreted.

8.13. Subtraction Soup

Based on the concept of the "personal fat threshold" developed by Professor Roy Taylor at the University of Newcastle upon Tyne, UK, probably the ultimate form of food reduction is a strict soup- and shake-based diet designed to reduce visceral fat in people suffering from type 2 diabetes [63]. Although hunger meant that strict adherence was a challenge, it was shown that insulin levels remained normal if the reduced body weight could be maintained over the long term [64]. However, the majority of "controlled-food" diets offer a less rigorous regime based around historical observations made in different parts of the world.

8.14. A Blue Diet?

Interestingly, as the dual inheritance hypothesis agrees with David Barker's 1990 article "The Fetal and Infant Origins of Adult Disease" in that the propensity for disease is set at birth [65], it follows that any person inheriting a fully functioning microbiome will tolerate an adequately nutritional diet that does not contain actual poisons. As an example, there are a series of so-called "Blue Zones" in which people tend to live long and healthy lives. While much effort focuses on what may be termed the "BZ Diet" [50] it is, perhaps, more likely that they owe their benefits to the circumstances of their birth, up to a century or so earlier. If so, then individuals subsequently inheriting a dysbiotic microbiome in more recent times may gain no advantage from such a diet.

8.15. A Tailored Diet

A different tack is taken with so-called personalised nutrition, which attempts to take account of an individual's genetic makeup, in addition to their intestinal bacteria [66]. However, this approach does have its detractors, specifically about terms related to glycaemic response, such as "normal postprandial" or "absolute and relative" [67]. Regardless of the exact definitions, however, a systematic review of moderate- to high-quality trials "did not show consistent benefits... in dietary, behavioural, or health outcomes" [68]. Interestingly, longer-term predictive studies looking at both dietary and inflammatory markers have uncovered a potential role for entities normally classed as parasites, such as *Blastocystis* species (protozoa often associated with detectable disease) [69].

8.16. Low-Carbohydrate Diets

One of the alternatives for the weight-gain energy balance model (EBM) summarised in Chapter 4 was the so-called carbohydrate–insulin model (CIM), championed by David Ludwig and his team. The idea is that the modern dietary environment exposes our endocrine systems to greater levels of carbohydrate than were found during evolution, leaving them effectively broken [70]. Whether or not this idea is correct (and it is very hard to obtain conclusive evidence as far as health and diet is concerned [71]), the low-carbohydrate diet philosophy is very successful, with the caveat that the term "low-carb" is hard to define exactly [72]. At the opposite end of the low-carbohydrate dietary spectrum, we have the so-called ketogenic diet. First developed to control epilepsy in children during the 1920s, this tasty-looking, very high-fat, moderate-protein diet sadly does not provide clear-cut answers when subjected to "an umbrella review of meta-analyses"; however, the authors suggest that longer-term, more comprehensive investigations may afford a more definitive answer [73]. Similar points apply to a recent pilot study noting benefits to both mental and metabolic health, as well as a reduction in symptoms relating to bipolar disorder and schizophrenia [74]. While we wait for confirmation of these results, however, it is clear that there are fewer of the theoretically good *bifidobacteria* in the intestine of people on the keto diet, along with a reduction in the levels of butyrate and other short-chain fatty acid-producing bacteria, although the implications of these findings for the health of individuals are not clear [75].

8.17. Omega-3 Fatty Acids

Of course, a relatively high-fat, low-carbohydrate diet allows an opportunity to increase the ratio of omega-3 polyunsaturated unsaturated fatty acids compared to the overwhelmingly more common omega-6 equivalents [76]. It seems reasonable to suppose that early humans were more reliant on marine food sources than our later farming ancestors, as the worldwide presence of extensive shell middens, dating from palaeolithic times, testifies [77]. Interestingly, the primary benefit of the omega-3 series of fatty acids is around birth, with supplementation being associated with many benefits to both mother and infant, especially involving brain structure and the immune system [78]. Indeed, it may well be that many of the benefits of the more traditional eating patterns, such as the so-called "Mediterranean diet", actually come from the greater levels of omega-3 polyunsaturated fatty acids that are available around pregnancy and infancy.

8.18. Olive Oil and the Mediterranean Diet

There are a number of diets, based on regions in which people seemed to achieve good levels of health and longevity, of which some variation of the Mediterranean diet is probably the best known [79], seemingly superior to low-fat diets [80]. Few subjects could have had more words written about them than the Mediterranean Diet, from its celebration as a holistic culture on the one hand [81] to virtual advertising for olive oil on the other [82]. Olive oil has itself undergone a parallel transition, firstly being lauded as a source of "healthy" monounsaturated fatty acids then, as the virtue of the extra-virgin designation gained a higher profile, attention shifted to the polyphenols in imparting its green colour and characteristic bitterness. In turn, these were assumed to be absorbed into the body as so-called antioxidants, at least until they were shown to interact with the gut microbiota [83]. On the whole, it is likely that the secret lies in the "cold pressing" process that yields an impure oil, in which the polyphenol "contaminants" have a complementary value to the oil itself.

8.19. Antioxidants or Polyphenols?

Interestingly, the "healthy" polyphenols were just being uncovered in the first decade of the 21st century, perhaps a little too early for the more modern understanding of the microbiome, and the seemingly reasonable assumption was made that they possess tocopherol-like antioxidant ability. However, initial enthusiasm about these suspected polyphenol "antioxidants" should have been dented when they were found to be of low bioavailability [84] and, even more significantly, phenols are normally toxic within the body [85]. However, such is the drive to find medicinal agents active against non-communicable disease that research continues unabated. Turmeric is a case in point [86]. While it is a natural product with a wide variety of components, much effort has been directed at increasing the bioavailability of one of them, curcumin, in spite of the fact that it is considered to be too reactive from a medicinal chemistry standpoint [87]. While there are an almost infinite number of natural products, and the search for the single magic ingredient continues, the polyphenols are most likely to exert their benefits through the microbiome, and are probably best obtained by eating the plant sources in which they are normally found.

8.20. Polyphenols: Stimulating the Microbiome

While many people will have heard of the benefits of dietary fibre, the term "polyphenol" is a relative newcomer. These compounds represent a class of aromatic compounds normally associated with the strong colours of fruits and leafy vegetables, as well as the scents of herbs and spices [88]. As described in the last paragraph, owing to an initial lack of knowledge of their mode of action, they are still often described by the default term "antioxidants", but they are now considered to be more likely to improve health via stimulation of the gut microbiota reducing residence time in the small intestine. Tannins, a complex form of mixed polyphenols that have been described as "anti-nutrients" due to their ability to entrap otherwise valuable metal ions [89], may be responsible for at least some of the beneficial effects of tea and coffee, for example [90].

8.21. Dietary Fibre: Feeding the Microbiota

The term dietary fibre represents a range of hard-to-digest complex carbohydrates such as the galactose-based oligosaccharides found in milk, or the fructose-based fructans normally associated with the roots and leaves of vegetables. Needless to say, as soon as natural products are identified, the search is on for therapeutic agents [91] and/or medicinal bioactivities [92]. These substances are the so-called prebiotics, normally associated with the production of short-chain fatty acids by bacteria within the intestine [93], of which perhaps the best known are the inulin fructans [94]. Interestingly, while no carbohydrates are technically considered to be "essential" in the way that certain fats and amino acids are, nowadays they are widely considered to be valuable for optimal health [95]. Indeed, although cooked starchy foods such as potatoes are normally readily digested, it is relatively easy to modify them by chilling to afford so-called resistant starch, thus slowing down their conversion to glucose and hence, in principle, lessening any glycaemic load on consumption [96]. In turn, this changes the ratios of gut bacteria reported to lead to weight loss [97], although it remains possible that any observed weight loss is actually due to a temporary strengthening of the gut–brain axis reducing transit time in the small intestine.

8.22. FODMAP Intolerance: Irritating the Bowel

Of course, microbiome-related disease is not restricted to slowed peristalsis and the accumulation of excess fat, as inflammation and autoimmune disease add to the burden, the intestine being especially susceptible to microbiome-related problems. Accordingly, it is not surprising that food is associated with bowel disorders—if not inflammatory, such as ulcerative colitis or coeliac disease [98], then irritable, causing painful spasms [99]. Patient experience pointed to a wide number of the so-called fermentable saccharides being involved, and eventually the term "FODMAP" was adopted, standing for "Fermentable Oligo-, Di-, Mono- saccharides, And Polyols"—the latter referring to a class of loosely related carbohydrate-like substances. Translated, this catchy acronym means that any carbohydrate and related substances may cause irritable bowel-like symptoms, nominally requiring a low-FODMAP diet to avoid such problems. Unfortunately, these conditions are very variable, and it is possible that the services of a nutritionist may, in itself, ease the condition, if only by stimulating the gut–brain axis in a fashion reminiscent of the placebo effect [26]. Needless to say, microbes are involved [100], but attempts to supplement them by bifidobacteria are not always successful [101]. Sadly, large, long-term randomised controlled studies are missing, while doubts about the cost-effectiveness of such low-FODMAP diets are growing, although there remain hopes that they may eventually contribute to the control of more serious conditions [102]. Although coeliac disease itself is known to be a potentially fatal allergy to gluten-related peptides [103], gluten intolerance, like FODMAP, tends to afford irritable bowel-like symptoms, leaving clinicians confused [99]. Sadly, the overall situation has not clarified in more recent years, with "wheat sensitivity" being added to gluten intolerance [104]. It seems feasible that there is a degree of subclinical inflammation of the gut wall, with dietary fibre-/gluten-stimulated bacterial growth causing a level of irritation, which can be detected by the patients themselves, but without yielding any interpatient disease-specific biomarkers. Accordingly, this is likely to be a "classic" microbiome-function deficiency disease: vulnerable to placebo and/or nocebo effects dependent upon the fluctuating strength of the gut–brain axis. Unfortunately, this diagnosis is not, in itself, helpful to the patient, nor is the fact that bowel discomforts are more likely to be associated with other microbiome-related conditions such as cardiovascular disease and/or poor mental health [105].

8.23. Lactose Intolerance

Another major component of breast milk is lactose, a disaccharide that is cleaved into more readily digestible monosaccharides by the enzyme lactase, an ability which almost all mammals lose after weaning. The exceptions are human populations who have spent many generations living a cattle-based lifestyle, consuming lactose-rich animal milk derivatives, in which lactase production has continued into adulthood. Accordingly, those people lacking the necessary enzyme who consume lactose will experience bacterial fermentation and may feel the gastrointestinal discomfort known as lactose intolerance [106]. It is important to note the accumulation of factors: the absence of lactase; the presence of microbiome-function deficiency; and the existence of another factor, such as intestinal inflammation, which leads to the actual abdominal discomfort itself. In the absence of the last factor, milk oligosaccharides have actually been considered as microbiome-stimulating therapeutic agents in humans [91].

8.24. C-Section Delivery: Losing the Microbes

Essentially the same arguments apply to the delivery of babies by caesarean section under sterile conditions as for the use of antibiotics, that key microbes may not be transferred from the maternal microbiome to the neonate gut. It is known that the neonate bacteriome will differ greatly between vaginally or C-section-delivered babies, and that a post-birth procedure, often called vaginal seeding, will erase some of the differences, but the health outcomes remain unclear [107]. On the whole, however, it seems sensible to count vaginal seeding as the default option, unless there are very specific reasons not to do so. It is worth noting that standard hospital procedures are not as sterile as the obtention of so-called germ-free, or gnotobiotic, animals for research purposes [108] and it may be possible for differential (beneficial) contamination to occur during the C-section delivery of twins, perhaps leading to correspondingly variable health outcomes. Interestingly, "twin studies" (see paragraph 4.11), the comparison of genetically identical and non-identical twins, have long been used to distinguish genetic from environmental influences, often without any consideration of the also-heritable microbiome [109]. It is important to note that essentially all reported studies refer exclusively to the bacteriome, whereas the dual inheritance hypothesis postulates the presence of "benegen" microeukaryotes, potentially as evolved partners of dendritic cells [110], or of other sentinel cells of mammalian immune systems [111]. Furthermore, the nature and epidemiology of non-communicable disease is best explained if, as seems likely, malfunctioning microbiomes are a consistent feature of industrialised countries and, granted that the roots of non-communicable disease can be traced back at least to the early 19th century [112], it seems that the single consistent factor over all this time will be toxic heavy metals distributed around the world. In turn, it is likely that such poisoning has led to a progressive loss of microbiome function, with damage accumulated over the generations and, perhaps, leaving a signature in the shape of "inherited" non-communicable disease, even in the absence of the initial pollutant. What is needed is a way to look inside the gut itself...

8.25. Ingestible Sensors

The miniaturisation of electronics has allowed the development of pill-like devices that are able to sense and transmit information relating to events inside the gut lumen [113]. The technology has been comprehensively reviewed [114], covered by patent landscape analysis [115], and is classed as "minimally invasive" for human study purposes [116]. The first working system to be commonly employed is based on lights and a camera, transmitting live images from the inside of the gut [117]. More recently, a pill with an in-built accelerometer has been developed and trialled to monitor vital signs such as heartbeat and breathing rate [118]. A chloride ion detector has been built inside a capsule, alongside a transmitter, so that exposure to gastric juices causes the device to emit a signal, designed to eventually reach key stakeholders such as research teams and healthcare providers [119]. Aside from chemical sensors, bacteria can be engineered to produce a luminescent signal on their detection of specific agents, in one case to detect a haemorrhage inside the stomach of a pig [120]. A more compact device is self-powered and is being developed specifically for the real-time measurement of metabolites within the gastrointestinal tract [121].

8.26. Assessing the Damage

It could be that the single most accessible measure of microbiome-related damage in the adult is the slowdown in intestinal transit time and, therefore, wireless motility capsules could be used more widely as a screening tool to estimate gut health [122], detecting the high-intestinal-motility traditional peoples as reported by Burkitt [49]. In particular, gut motility in populations that are not experiencing high levels of non-communicable conditions, e.g., the high-microbial-diversity Tanzanian Hadza, could be most enlightening [48]. As discussed above, checking gut motility is only one aspect of a wider range of opportunities afforded by the use of ingestible sensors. Not only is work underway on the faecal metabolome, the sum total of all small molecules contributed by food substances, gut wall, and microbial metabolites but, most importantly, the rate of change with respect to the level of industrialisation is also being assessed [123]. Their samples were taken from across the world, including the USA, Burkina Faso, and Peru, and can be compared to the (bacterial) microbiome data being accumulated within "urbanising" Brazilian Amazonian populations [124]. In fact, it seems that much of the work required for the determination of semiochemical action is already well underway, with the prime difficulty likely to be the recruitment of subjects from non-polluted parts of the world not suffering from extensive non-communicable disease. Fortunately, animal studies will be helpful, perhaps using feral pigs or similar undomesticated large animals. In principle, the wild baboons of the Kenyan Amboseli research project would be the perfect primate subject—if these intelligent animals could be persuaded to eat non-food-like objects, that is [125]. The ultimate aim, of course, is to improve the health of future generations by understanding the role of the fully functioning microbiome.

8.27. Summary: Human Biology

By this stage in the book, the reader may be reconsidering some assumptions regarding human biology. For example, which represents the "standard human value" for gut motility: people living the "traditional" or "modern, Western" lifestyles as noted by Denis Burkitt [49]? Indeed, it seems feasible that a poor microbial inheritance is the primary cause for obesity and other non-communicable diseases related to the circulatory system. As a consequence, it may be that the least affected may remain healthy on one of the more "balanced" diets, such as the so-called Mediterranean diet, for example [79]. By contrast, those more seriously affected may require more direct pharmacological intervention, such as metformin or statins, working well beyond their original indications of managing diabetes [31] or lowering blood cholesterol [126], respectively. Of course, the GLP-1 receptor agonists effectively enforce dieting by bypassing normal appetite regulation, albeit with further lengthening of gut motility [44], while it is encouraging to note that statin use seems to lower cancer mortality [127]. Finally, osteoarthritis is an example of a disease in which experienced pain is not closely related to the objective assessment of damage. Accordingly, it may be possible to leverage the placebo effect to advantage, even while making clinical trials more difficult [128]. In summary, while it seems ridiculous to assert that the same mechanism, dysbiosis, both gives rise to the disease and allows its amelioration, at this stage we simply do not know.

References

1. Éliás, J.E.; Barna, V.; Patoni, C.; Demeter, D.; Veres, D.S.; Bunduc, S.; Eröss, B.; Hegyi, P.; Földvári-Nagy, L.; Lenti, K. Probiotic supplementation during antibiotic treatment is unjustified in maintaining the gut microbiome diversity: A systematic review and meta-analysis. *BMC Med.* **2023**, *21*, 262. [CrossRef] [PubMed]
2. Konstantinidis, T.; Tsigalou, C.; Karvelas, A.; Stavropoulou, E.; Voidarou, C.; Bezirtzoglou, E. Effects of antibiotics upon the gut microbiome: A review of the literature. *Biomedicines* **2020**, *8*, 502. [CrossRef] [PubMed]
3. Druvefors, E.; Myrelid, P.; Andersson, R.E.; Landerholm, K. Female and male fertility after colectomy and reconstructive surgery in inflammatory bowel disease: A national cohort study from Sweden. *J. Crohn's Colitis* **2023**, *17*, 1631–1638. [CrossRef] [PubMed]
4. Baggio, S.; Pomini, P.; Zecchin, A.; Garzon, S.; Bonin, C.; Santi, L.; Festi, A.; Franchi, M.P. Delivery and pregnancy outcome in women with bowel resection for deep endometriosis: A retrospective cohort study. *Gynecol. Surg.* **2015**, *12*, 279–285. [CrossRef]
5. da Cunha, B.R.; Fonseca, L.P.; Calado, C.R.C. Antibiotic discovery: Where have we come from, where do we go? *Antibiotics* **2019**, *8*, 45. [CrossRef] [PubMed]
6. Rhea, M.C.; Dobson, A.; O'Sullivan, O.; Crispie, F.; Fouhy, F.; Cotter, P.D.; Shanahan, F.; Kiely, B.; Hill, C.; Ross, R.P. Effect of broad- and narrow-spectrum antibiotics on *Clostridium difficile* and microbial diversity in a model of the distal colon. *Proc. Natl. Acad. Sci. USA* **2011**, *108*, 4639–4644. [CrossRef] [PubMed]
7. Wilhelm, S.W.; Suttle, C.A. Viruses and nutrient cycles in the sea. *BioScience* **1999**, *49*, 781–788. [CrossRef]
8. Babakhanova, A.T.; Dzhumabekhov, A.T.; Zhao, A.V.; Kuandykov, Y.K.; Tanabayeva, S.B.; Fakhradiyev, I.R.; Nazarenko, Y.; Saliev, T.M. Impact of appendectomy on gut microbiota. *Surg. Infect.* **2021**, *22*, 651–661. [CrossRef] [PubMed]
9. Aranaga, C.; Pantoja, L.D.; Martínez, E.A.; Falco, A. Phage therapy in the era of multidrug resistance in bacteria: A systematic review. *Int. J. Mol. Sci*, **2022**, *23*, 4577. [CrossRef] [PubMed]
10. Low, C.X.; Tan, L.T.-H.; Ab Mutalib, N.-S.; Pusparajah, P.; Goh, B.-H.; Chan, K.-G.; Letchumanan, V.; Lee, L.-H. Unveiling the impact of antibiotics and alternative methods for animal husbandry: A review. *Antibiotics* **2021**, *10*, 578. [CrossRef] [PubMed]
11. Gaskins, H.R.; Collier, C.T.; Anderson, D.B. Antibiotics as growth promotants: Mode of action. *Anim. Anim. Biotechnol.* **2002**, *13*, 29–42. [CrossRef] [PubMed]
12. Marsella, R. Atopic dermatitis in domestic animals: What our current understanding is and how this applies to clinical practise. *Vet. Sci.* **2021**, *8*, 124. [CrossRef] [PubMed]

13. Blaser, M.J.; Falkow, S. What are the consequences of the disappearing human microbiota? *Nat. Rev. Microbiol.* **2009**, *7*, 887–894. [CrossRef] [PubMed]
14. Stambler, I. Elie Metchnikov-The founder of longevity science and a founder of modern medicine: In honour of the 170th anniversary. *Adv. Gerontol.* **2015**, *5*, 201–208. [CrossRef]
15. Swanson, K.S.; Gibson, G.R.; Hutkins, R.; Reimer, R.A.; Reid, G.; Verbeke, K.; Scott, K.P.; Holscher, H.D.; Azad, M.B.; Delzenne, N.M.; et al. The international scientific association for probiotics and prebiotics (ISAPP) consensus statement on the definition and scope of synbiotics. *Nat. Rev. Gastroenterol. Hepatol.* **2020**, *17*, 687–701. [CrossRef] [PubMed]
16. Ma, L.; Tu, H.; Chen, T. Postbiotics in human health: A narrative review. *Nutrients* **2023**, *15*, 291. [CrossRef] [PubMed]
17. Zagórska, A.; Marcinkowska, M.; Jamrozik, M.; Wiśniowska, B.; Paśko, P. From probiotics to psychobiotics—The gut-brain axis in psychiatric disorders. *Benef. Microbes* **2020**, *11*, 717–732. [CrossRef] [PubMed]
18. Lee, N.; Park, Y.S.; Kang, D.K.; Paik, H.D. Paraprobiotics: Definition, manufacturing methods, and functionality. *Food Sci. Biotechnol.* **2023**, *32*, 1981–1991. [CrossRef] [PubMed]
19. Griffith, F. The significance of pneumococcal types. *J. Hyg.* **1928**, *27*, 113–159. [CrossRef] [PubMed]
20. Kerksick, C.M.; Moon, J.M.; Jäger, R. It's dead! Can postbiotics really help performance and recovery? A systematic review. *Nutrients* **2024**, *16*, 720. [CrossRef] [PubMed]
21. McGovern, P.E.; Zhang, J.; Tang, J.; Zhang, Z.; Hall, G.R.; Moreau, R.A.; Nuñez, A.; Butrym, E.D.; Richards, M.P.; Wang, C.S.; et al. Fermented beverages of pre- and proto-historic China. *Proc. Natl. Acad. Sci. USA* **2004**, *101*, 17593–17598. [CrossRef] [PubMed]
22. Mannaa, M.; Han, G.; Seo, Y.-S.; Park, I. Evolution of food fermentation processes and the use of multi-omics in deciphering the roles of the microbiota. *Foods* **2021**, *10*, 2861. [CrossRef] [PubMed]
23. Liu, S.; Hu, J.; Zhong, Y.; Hu, X.; Yin, J.; Xiong, T.; Nie, S.; Xie, M. A review: Effects of microbial fermentation on the structure and bioactivity of polysaccharides in plant-based foods. *Food Chem.* **2024**, *440*, 137453. [CrossRef] [PubMed]
24. Rizo, J.; Guillén, D.; Díaz-Ruiz, G.; Wacher, C.; Encarnación, S.; Sánchez, S.; Rodríguez-Sanoja, R. Metaproteomic insights into the microbial community in pozol. *Front. Nutr.* **2021**, *8*, 714814. [CrossRef] [PubMed]
25. Dahl, W.J.; Agro, N.C.; Eliasson, A.M.; Mialki, K.L.; Olivera, J.D.; Rusch, N.C.; Young, C. Heath benefits of fiber fermentation. *J. Am. Coll. Nutr.* **2017**, *36*, 127–136. [CrossRef] [PubMed]
26. de Craen, A.J.; Kaptchuck, T.J.; Tijssen, J.G.; Kleijnen, J. Placebos and placebo effects in medicine: Historical overview. *J. R. Soc. Med.* **1999**, *92*, 511–515. [CrossRef] [PubMed]
27. Giddings, S.L.; Stevens, A.M.; Leung, D.T. Traveler's diarrhea. *Med. Clin. N. Am.* **2016**, *100*, 317–330. [CrossRef] [PubMed]
28. Drucker, D.J.; Habener, J.F.; Holst, J.J. Discovery, characterisation, and clinical development of the glucagon-like peptides. *J. Clin. Investig.* **2017**, *127*, 4217–4227. [CrossRef] [PubMed]

29. Mozaffarian, D. Perspective: Obesity-an unexplained epidemic. *Am. J. Clin. Nutr.* **2022**, *115*, 1445–1450. [CrossRef] [PubMed]
30. Bailey, C.J. Metformin: Historical overview. *Diabetologia* **2017**, *60*, 1566–1576. [CrossRef] [PubMed]
31. Dutta, S.; Shah, R.B.; Singhal, S.; Dutta, S.B.; Bansal, S.; Sinha, S.; Haque, M. Metformin: A review of potential mechanism and therapeutic utility beyond diabetes. *Drug Des. Dev. Ther.* **2023**, *17*, 1907–1932. [CrossRef] [PubMed]
32. Zhu, Y.; Song, G. Molecular origin and biological effects of exercise mimics. *J. Exerc. Sci. Fit.* **2024**, *22*, 73–85. [CrossRef] [PubMed]
33. Hofer, S.J.; Davinelli, S.; Bergmann, M.; Scapagnini, G.; Madeo, F. Calorie restriction mimetics in nutrition and clinical trials. *Front. Nutr.* **2021**, *8*, 717343. [CrossRef] [PubMed]
34. Rezabakhsh, A.; Mahmoodpoor, A.; Soleimanpour, H. Historical perspective of aspirin: A journey from discovery to clinical practise Ancient and modern history. *J. Cardiovasc. Thorac. Res.* **2021**, *13*, 179–180. [CrossRef] [PubMed]
35. Jóhannesson, T. Aspirin. Acetylsalicylic acid and aspirinlike drugs. A review. *Laeknabladid* **2000**, *86*, 755–768. [PubMed]
36. Elwood, P.; Morgan, G.; Watkins, J.; Protty, M.; Mason, M.; Adams, R.; Dolwani, S.; Pickering, J.; Delon, C.; Longley, M. Aspirin and cancer treatment: Systematic reviews and meta-analyses of evidence: For and against. *Br. J. Cancer* **2024**, *130*, 3–8. [CrossRef] [PubMed]
37. Hall, K.D.; Heymsfield, S.B.; Kemnitz, J.W.; Klein, S.; Schoeller, D.A.; Speakman, J.R. Energy balance and its components: Implications for body weight regulation. *Am. J. Clin. Nutr.* **2012**, *95*, 989–994. [CrossRef] [PubMed]
38. Careau, V.; Halsey, L.G.; Pontzer, H.; Ainslie, P.N.; Andersen, L.F.; Anderson, L.J.; Arab, L.; Baddou, I.; Bedu-Addo, K.; Blaak, E.E.; et al. Energy compensation and adiposity in humans. *Curr. Biol.* **2021**, *31*, 1–8. [CrossRef] [PubMed]
39. Robinson, E.; Stensel, D. Does physical activity cause weight loss? *Int. J. Obes.* **2023**, *47*, 91–92. [CrossRef] [PubMed]
40. Jensen, S.B.K.; Blond, M.B.; Sandsdal, R.M.; Olsen, L.M.; Juhl, C.R.; Lundgren, J.R.; Janus, C.; Stallknecht, B.M.; Holst, J.J.; Madsbad, S.; et al. Healthy weight loss maintenance with exercise, GLP-1 receptor agonist, or both combined followed by one year without treatment: A post-treatment analysis of a randomised placebo-controlled trial. *eClinicalMedicine* **2024**, *69*, 102475. [CrossRef] [PubMed]
41. Verhoeven, J.E.; Han, L.K.M.; Milligen, B.A.L.; Hu, M.X.; Révész, D.; Hoogendoorn, A.W.; Batelaan, N.M.; van Schalk, D.J.F.; van Balkam, A.J.L.M.; van Oppen, P.; et al. Antidepressants or running therapy: Comparing effects on mental and physical health in patients with depression and anxiety disorders. *J. Affect. Disord.* **2023**, *329*, 19–29. [CrossRef] [PubMed]
42. Chow, L.S.; Gerszten, R.E.; Pederson, B.K.; van Praag, H.; Trappe, S.; Febbraio, M.A.; Galis, Z.S.; Gao, Y.; Haus, J.M.; Lanza, I.R.; et al. Exerkines in health, resilience and disease. *Nat. Rev. Endocrinol.* **2022**, *18*, 273–289. [CrossRef] [PubMed]
43. Lavery, J.A.; Boutros, P.C.; Knight, D.; Tammela, T.; Moskovitz, C.S.; Jones, L.W. Association of exercise with pan-cancer incidence and overall survival. *Cancer Cell* **2024**, *42*, 169–171. [CrossRef] [PubMed]

44. Wang, L.; Wang, W.; Kaelber, D.C.; Xu, R.; Berger, N.A. GLP-1 receptor agonists and colorectal cancer risk in drug-naive patients with type 2 diabetes, with and without overweight/obesity. *JAMA Oncol.* **2024**, *10*, 256–258. [CrossRef] [PubMed]
45. Foster, R.; Lunn, J. 40th Anniversary briefing paper: Food availability and our changing diet. *Nutr. Bull.* **2007**, *32*, 187–249. [CrossRef]
46. Protsiv, M.; Ley, C.; Lankester, J.; Hastie, T.; Parsonnet, J. Decreasing human body temperature in the United States since the Industrial Revolution. *eLife* **2020**, *9*, e49555. [CrossRef] [PubMed]
47. Sandercock, G.R.H.; Cohen, D.D. Temporal trends in muscular fitness of English 10-year-olds 1998–2014: An allometric approach. *J. Sci. Med. Sport.* **2019**, *22*, 201–205. [CrossRef] [PubMed]
48. Carter, M.M.; Olm, M.R.; Merrill, B.D.; Dahan, D.; Tripathi, S.; Spencer, S.P.; Yu, F.B.; Jain, S.; Neff, N.; Jha, A.R.; et al. Ultra-deep sequencing of Hadza hunter-gatherers recovers vanishing gut microbes. *Cell* **2023**, *186*, 3111–3124. [CrossRef] [PubMed]
49. Burkitt, D.P. Some diseases characteristic of modern western civilization. *Br. Med. J.* **1973**, *1*, 274–278. [CrossRef] [PubMed]
50. Pes, G.M.; Dore, M.P.; Tsofliou, F.; Poulain, M. Diet and longevity in the Blue Zones: A set-and-forget issue? *Maturitas* **2022**, *164*, 31–37. [CrossRef] [PubMed]
51. Bostock, J. Of the catarrhus aestivus or summer catarrh. *Med. Chir. Trans.* **1828**, *14*, 437–446. [CrossRef] [PubMed]
52. Eknoyan, G. Adolphe Quetelet (1796–1874)-the average man and indices of obesity. *Nephrol. Dial. Transplant.* **2008**, *23*, 47–51. [CrossRef] [PubMed]
53. Sommer, I.; Teufer, B.; Szelag, M.; Nussbaumer-Streit, B.; Titscher, V.; Klerings, I.; Gartlehner, G. The performance of anthropometric tools to determine obesity: A systematic review and meta-analysis. *Sci. Rep.* **2020**, *10*, 12699. [CrossRef] [PubMed]
54. Koyama, T. New anthropometric measurement alternatives to BMI and hypertension. *Hypertens. Res.* **2023**, *46*, 776–777. [CrossRef] [PubMed]
55. St-Onge, M.P.; Wang, J.; Shen, W.; Wang, Z.; Allison, D.B.; Heshka, S.; Pierson, R.N.; Heymsfield, S.B. Dual-energy X-ray absorptiometry-measured lean soft tissue mass: Differing relation to body cell mass throughout the adult life span. *J. Gerontol. A Biol. Sci. Med. Sci.* **2004**, *59*, 796–800. [CrossRef] [PubMed]
56. Agbaje, A.O. Waist-circumference to height ratio had better longitudinal agreement with DEXA-measured fat mass than BMI in 7237 children. *Pediatr. Res.* **2024**, *96*, 1369–1380. [CrossRef] [PubMed]
57. Wise, J. Advise adults to keep waist size to less than half their height, says NICE. *BMJ* **2022**, *377*, o933. [CrossRef] [PubMed]
58. López-Gil, J.F.; García-Hermoso, A.; Smith, L.; Firth, J.; Trott, M.; Mesas, A.E.; Jiménez-López, E.; Gutiérrez-Espinoza, H.; Tárraga- López, P.J.; Victoria-Montesinos, D. Global proportion of disordered eating in children and adolescents: A systematic review and meta-analysis. *JAMA Pediatr.* **2023**, *177*, 363–372. [CrossRef] [PubMed]
59. Puckett, L.; Grayeb, D.; Khatri, V.; Cass, K.; Mehler, P. A comprehensive review of complications and new findings associated with anorexia nervosa. *J. Clin. Med.* **2021**, *10*, 2555. [CrossRef] [PubMed]

60. Elizabeth, L.; Machado, P.; Zinöcker, M.; Baker, P.; Lawrence, M. Ultra-processed foods and health outcomes: A narrative review. *Nutrients* **2020**, *12*, 1955. [CrossRef] [PubMed]
61. Doyle, J.; Alsan, M.; Skelley, N.; Lu, Y.; Cawley, J. Effect of an intensive food-as-medicine program on health and healthcare use. *JAMA Intern. Med.* **2024**, *184*, 154–163. [CrossRef] [PubMed]
62. Hall, K.D.; Ayuketah, A.; Brychta, R.; Cai, H.; Cassimatis, T.; Chen, K.Y.; Chung, S.T.; Costa, E.; Courville, A.; Darcey, V.; et al. Ultra-processed diets cause excess calorie intake and weight gain: An inpatient randomized controlled trial of ad libitum food intake. *Cell Metab.* **2019**, *30*, 67–77.e3. [CrossRef] [PubMed]
63. Taylor, R.; Holman, R.R. Normal weight individuals who develop type 2 diabetes: The personal fat threshold. *Clin. Sci.* **2015**, *128*, 405–410. [CrossRef] [PubMed]
64. Taylor, R. Calorie restriction for long-term remission of type 2 diabetes. *Clin. Med.* **2019**, *19*, 37–42. [CrossRef] [PubMed]
65. Barker, D.J. The fetal and infant origins of adult disease. *BMJ* **1990**, *301*, 1111. [CrossRef] [PubMed]
66. Zeevi, D.; Korem, T.; Zmora, N.; Israeli, D.; Rothschild, D.; Weinberger, A.; Ben-Yakov, O.; Lador, D.; Avnit-Sagi, T.; Lotan-Pompan, M.; et al. Personalized nutrition by prediction of glycaemic responses. *Cell* **2015**, *163*, 1079–1094. [CrossRef] [PubMed]
67. Wolever, T. Personalized nutrition by prediction of glycaemic responses: Fact or fantasy? *Eur. J. Clin. Nutr.* **2016**, *70*, 411–413. [CrossRef] [PubMed]
68. Shyam, S.; Lee, K.X.; Tan, A.S.W.; Khoo, T.A.; Harikrishnan, S.; Lalani, S.A.; Ramadas, A. Effect of personalised nutrition on dietary, physical activity, and health outcomes: A systematic review of randomized trials. *Nutrients* **2022**, *14*, 4104. [CrossRef] [PubMed]
69. Asnicar, F.; Berry, S.E.; Valdes, A.M.; Nguyen, L.H.; Piccinno, G.; Drew, D.A.; Leeming, E.; Gibson, R.; Le Roy, C.; Al Khatib, H.; et al. Microbiome connections with host metabolism and habitual diet from 1,098 deeply phenotyped individuals. *Nat. Med.* **2021**, *27*, 321–332. [CrossRef] [PubMed]
70. Ludwig, D.S.; Aronne, L.J.; Astrup, A.; de Cabo, R.; Cantley, L.C.; Friedman, M.I.; Heymsfield, S.B.; Johnson, J.D.; King, J.C.; Krauss, R.M.; et al. The carbohydrate-insulin model: A physiological perspective on the obesity pandemic. *Am. J. Clin. Nutr.* **2021**, *114*, 1873–1885. [CrossRef] [PubMed]
71. Hall, K.D. A review of the carbohydrate-insulin model of obesity. *Eur. J. Clin. Nutr.* **2017**, *71*, 323–326. [CrossRef] [PubMed]
72. Chacón, V.; Cara, K.C.; Chung, M.; Taylor, C.W. Defining "low-carb" in the scientific literature: A scoping review of clinical studies. *Crit. Rev. Food Sci. Nutr.* **2024**, *65*, 1792–1801. [CrossRef] [PubMed]
73. Chen, S.; Su, X.; Feng, Y.; Li, R.; Liao, M.; Fan, L.; Liu, J.; Chen, S.; Zhang, S.; Cai, J.; et al. Ketogenic diet and multiple health outcomes: An umbrella view of meta-analyses. *Nutrients* **2023**, *15*, 4161. [CrossRef] [PubMed]
74. Sethi, S.; Wakeham, D.; Ketter, T.; Hooshmand, F.; Bjornstad, J.; Richards, B.; Westman, E.; Krauss, R.M.; Saslow, L. Ketogenic diet intervention on metabolic and psychiatric health in bipolar and schizophrenia: A pilot trial. *Psychiatry Res.* **2024**, *335*, 115866. [CrossRef] [PubMed]

75. Rew, L.; Harris, M.D.; Goldie, J. The ketogenic diet: Its impact on human gut microbiota and potential consequent health outcomes: A systematic literature review. *Gastroenterol. Hepatol. Bed Bench* **2022**, *15*, 326–342. [PubMed]
76. Simopoulos, A.P. *The Healthiest Diet for You: Scientific Aspects*; MDPI: Basel, Switzerland, 2022.
77. Rick, T.C. Shell midden archaeology: Current trends and future directions. *J. Archaeol. Res.* **2023**, *32*, 309–366. [CrossRef]
78. Firouzabadi, F.D.; Shab-Bidar, S.; Jayedi, A. The effects of omega-3 polyunsaturated fatty acids supplementation in pregnancy, lactation, and infancy: An umbrella review of meta-analyses of randomized trials. *Pharmacol. Res.* **2022**, *177*, 106100. [CrossRef] [PubMed]
79. Davis, C.; Bryan, J.; Hodgson, J.; Murphy, K. Definition of the Mediterranean diet: A literature review. *Nutrients* **2015**, *7*, 9139–9153. [CrossRef] [PubMed]
80. Delgado-Lista, J.; Alcala-Diaz, J.F.; Torres-Peña, J.D.; Quintana-Navarro, G.M.; Fuentes, F.; Garcia-Rios, A.; Ortiz-Morales, A.M.; Gonzalez-Requero, A.I.; Perez-Caballero, A.I.; Yubero-Serrano, E.M.; et al. Long-term secondary prevention of cardiovascular disease with a Mediterranean diet and a low-fat diet (CORDIOPREV): A randomised controlled trial. *Lancet* **2022**, *390*, 1876–1885. [CrossRef] [PubMed]
81. Lacatusu, C.-M.; Grigorescu, E.-D.; Floria, M.; Onofriescu, A.; Mihai, B.-M. The Mediterranean diet: From an environment-driven food culture to an emerging medical prescription. *Int. J. Environ. Res. Public Health* **2019**, *16*, 942. [CrossRef] [PubMed]
82. Finicelli, M.; Squillaro, T.; Galderisi, U.; Peluso, G. Polyphenols, the healthy brand of olive oil: Insights and perspectives. *Nutrients* **2021**, *13*, 3831. [CrossRef] [PubMed]
83. Millman, J.; Okamoto, S.; Teruya, T.; Uema, T.; Ikematsu, S.; Shimabukuro, M.; Masuzaki, H. Extra-virgin olive oil and the gut-brain axis: Influence on gut microbiota, mucosal immunity, and cardiometabolic and cognitive health. *Nutr. Rev.* **2021**, *79*, 1362–1374. [CrossRef] [PubMed]
84. D'Archivio, M.; Filesi, C.; Vari, R.; Scazzocchio, B.; Masella, R. Bioavailability of the polyphenols: Status and controversies. *Int. J. Mol. Sci.* **2010**, *11*, 1321–1342. [CrossRef] [PubMed]
85. Shadnia, H.; Wright, J.S. Understanding the toxicity of phenols: Using quantitative structure-activity relationships and enthalpy changes to discriminate between possible mechanisms. *Chem. Res. Toxicol.* **2008**, *21*, 1197–1204. [CrossRef] [PubMed]
86. Rolfe, V.; Mackonochie, M.; Mills, S.; MacLennon, E. Turmeric / curcumin and health outcomes, A meta-review of systematic reviews. *Eur. J. Integr. Med.* **2020**, *40*, 101252. [CrossRef]
87. Nelson, K.M.; Dahlin, J.L.; Bisson, J.; Graham, J.; Pauli, G.F.; Walters, M.A. The essential medicinal chemistry of curcumin. *J. Med. Chem.* **2017**, *60*, 1620–1637. [CrossRef] [PubMed]
88. Garza, M.C.; Pérez-Calahorra, S.; Rodrigo-Carbò, C.; Sánchez-Calavera, M.A.; Jarauta, E.; Mateo-Gallego, R.; Garcia-Rubio, I.; Limiquiz-Maneo, I. Effect of herbs and spices present in the Mediterranean diet on the glycemic profile in type 2 diabetes subjects: A systematic review and meta-analysis. *Nutrients* **2024**, *16*, 756. [CrossRef] [PubMed]

89. Sharma, K.; Kumar, V.; Kaur, J.; Tanwar, B.; Goyal, A.; Sharma, R.; Gat, Y.; Kumar, A. Health effects, sources, utilization and safety of tannins: A critical review. *Toxin Rev.* **2019**, *40*, 432–444. [CrossRef]
90. Zhang, Y.; Yang, H.; Li, S.; Li, W.-D.; Wang, Y. Consumption of coffee and tea and risk of developing stroke, dementia, and poststroke dementia: A cohort study in the UK Biobank. *PLoS Med.* **2021**, *18*, e1003830. [CrossRef] [PubMed]
91. Rousseaux, A.; Brosseau, C.; le Gall, S.; Piloquet, H.; Barbarot, S.; Bodinier, M. Human milk oligosaccharides: Their effects on the host and their potential as therapeutic agents. *Front. Immunol.* **2021**, *12*, 680911. [CrossRef] [PubMed]
92. Wang, M.; Cheong, K.-L. Preparation, structural characterisation, and bioactivities of fructans: A review. *Molecules* **2023**, *28*, 1613. [CrossRef] [PubMed]
93. Fusco, W.; Lorenzo, M.B.; Cintoni, M.; Porcari, S.; Rinninela, E.; Kaitsas, F.; Lener, E.; Mele, M.C.; Gasbarrini, A.; Collado, M.C.; et al. Short-chain fatty-acid-producing bacteria: Key components of the human gut microbiota. *Nutrients* **2023**, *15*, 2211. [CrossRef] [PubMed]
94. Hughes, R.L.; Alvarado, D.A.; Swanson, K.S.; Holscher, H.D. The prebiotic potential of inulin-type fructans: A systematic review. *Adv. Nutr.* **2022**, *13*, 492–529. [CrossRef] [PubMed]
95. McKeown, N.; Fahey, G.C., Jr.; Slavin, J.; van der Kamp, J.-W. Fibre intake for optimal health: How can healthcare professionals support people to reach dietary recommendations? *BMJ* **2022**, *378*, e054370. [CrossRef] [PubMed]
96. Robertson, T.M.; Brown, J.E.; Fielding, B.A.; Hovorka, R.; Robertson, M.D. Resistant starch production and glucose release from pre-prepared chilled food: The SPUD project. *Nutr. Bull.* **2021**, *46*, 52–59. [CrossRef] [PubMed]
97. Li, H.; Zhang, L.; Li, J.; Wu, Q.; Qian, L.; He, J.; Ni, Y.; Kovatcheva-Datchary, P.; Yuan, R.; Liu, S.; et al. Resistant starch facilitates weight loss in humans by reshaping the gut microbiota. *Nat. Metab.* **2024**, *6*, 578–597. [CrossRef] [PubMed]
98. Kleinjans, M.; Schneider, C.V.; Bruns, T.; Strnad, P. Phenome of coeliac disease vs. inflammatory bowel disease. *Sci. Rep.* **2022**, *12*, 14572. [CrossRef] [PubMed]
99. Makharia, A.; Catassi, C.; Makharia, G.K. The overlap between irritable bowel syndrome and non-celiac gluten sensitivity: A clinical dilemma. *Nutrients* **2015**, *7*, 10417–10426. [CrossRef] [PubMed]
100. Vandeputte, D.; Joossens, M. Effects of low and high FODMAP diets on human gastrointestinal microbiota composition in adults with intestinal diseases: A systematic review. *Microorganisms* **2020**, *8*, 1638. [CrossRef] [PubMed]
101. Pratt, C.; Campbell, M.D. The effect of bifidobacterium on reducing symptomatic pain in patients with irritable bowel syndrome: A systematic review. *Probiotics Antimicrob Proteins* **2020**, *12*, 834–839. [CrossRef] [PubMed]
102. Bellini, M.; Tonarelli, S.; Nagy, A.T.; Pancetti, A.; Costa, F.; Ricchiuti, A.; de Bortoli, N.; Mosca, M.; Marchi, S.; Rossi, A. Low FODMAP diet: Evidence, doubts, and hopes. *Nutrients* **2020**, *12*, 148. [CrossRef] [PubMed]
103. Silano, M.; Vincentini, O.; De Vincenzi, M. Toxic, immunostimulatory and antagonist gluten peptides in celiac disease. *Curr. Med. Chem.* **2009**, *16*, 1489–1498. [CrossRef] [PubMed]

104. Sergi, C.; Villanacci, V.; Carroccio, A. Non-celiac wheat sensitivity: Rationality and irrationality of a gluten-free diet in individuals affected with non-celiac disease: A review. *BMC Gastroenterol.* **2021**, *21*, 5. [CrossRef] [PubMed]
105. Drossman, D.A. Functional gastrointestinal disorders: History, pathophysiology, clinical features and Rome IV. *Gastroenterology* **2016**, *150*, 1262–1279. [CrossRef] [PubMed]
106. Wiley, A.S. Lactose intolerance. *Evol. Med. Public Health* **2020**, *2020*, 47–48. [CrossRef] [PubMed]
107. Ríos-Covian, D.; Langella, P.; Martín, R. From short- to long-term effects of C-section delivery on microbiome establishment and host health. *Microorganisms* **2021**, *9*, 2122. [CrossRef] [PubMed]
108. Basic, M.; Bleich, A. Gnotobiotics: Past, present and future. *Lab. Anim.* **2019**, *53*, 232–243. [CrossRef] [PubMed]
109. Hagenbeek, F.A.; Hirzinger, J.S.; Breunig, S.; Bruins, S.; Kuznetsov, D.V.; Schut, K.; Odintsova, V.V.; Boomsma, D.I. Maximising the value of twin studies in health and behaviour. *Nat. Hum. Behav.* **2023**, *7*, 849–860. [CrossRef] [PubMed]
110. Banchereau, J.; Briere, F.; Caux, C.; Davoust, J.; Lebecque, S.; Liu, Y.-J.; Pulendran, B.; Palucka, K. Immunobiology of dendritic cells. *Annu. Rev. Immunol.* **2000**, *18*, 767–811. [CrossRef] [PubMed]
111. Delgado, M.-G.; Lennon-Duménil, A.-M. How cell migration helps immune cells. *Front. Cell Dev. Biol.* **2022**, *10*, 932472. [CrossRef] [PubMed]
112. Bostock, J. Case of a periodical affection of the eyes and chest. *Med. Chir. Trans.* **1819**, *10*, 161–165. [CrossRef] [PubMed]
113. Kalantar-Zadeh, K.; Ha, N.; Ou, J.Z.; Berean, K.J. Ingestible sensors. *ACS Sens.* **2017**, *2*, 468–483. [CrossRef] [PubMed]
114. Wu, Y.; Ye, D.; Shan, Y.; He, S.; Su, Z.; Liang, J.; Zheng, J.; Yang, Z.; Yang, H.; Xu, W.; et al. Edible and nutritive electronics: Materials, fabrications, components, and applications. *Adv. Mater. Technol.* **2020**, *5*, 2000100. [CrossRef]
115. Litvinova, O.; Klager, E.; Tzvetkov, N.T.; Kimberger, O.; Kletecka-Pulker, M.; Willschke, H.; Atanasov, A.G. Digital pills with ingestible sensors: Patent landscape analysis. *Pharmaceuticals* **2022**, *15*, 1025. [CrossRef] [PubMed]
116. Beardslee, L.A.; Banis, G.E.; Chu, S.; Liu, S.; Chapin, A.A.; Stine, J.M.; Pasricha, P.J.; Ghodossi, R. Ingestible sensors and sensing systems for minimally invasive diagnosis and monitoring: The next frontier in minimally invasive screening. *ACS Sens.* **2020**, *5*, 891–910. [CrossRef] [PubMed]
117. Wang, A.; Banerjee, S.; Barth, B.A.; Bhat, Y.M.; Chauhan, S.; Gottlieb, K.T.; Konda, V.; Maple, J.T.; Murad, F.; Plau, P.R.; et al. Wireless capsule endoscopy. *Gastrointest. Endosc.* **2013**, *78*, 805–815. [CrossRef] [PubMed]
118. Traverso, G.; Finomore, V., Jr.; Mahoney, J., III; Kupec, J.; Stansbury, R.; Bacher, D.; Pless, B.; Schuetz, S.; Hayward, A.; Langer, R.; et al. First-in-human trial of an ingestible vitals-monitoring pill. *Device* **2023**, *1*, 100125. [CrossRef]
119. Chai, P.R.; Vaz, C.; Goodman, G.R.; Albrechta, H.; Huang, H.; Rosen, R.K.; Boyer, E.W.; Mayer, K.H.; O'Cleirigh, C. Ingestible electronic sensors to measure instantaneous medication adherence: A narrative review. *Digit. Health* **2022**, *8*, 20552076221083119. [CrossRef] [PubMed]

120. Mimee, M.; Nadeau, P.; Hayward, A.; Carim, S.; Flanagan, S.; Jerger, L.; Collins, J.; McDonnell, S.; Swartwout, R.; Citorik, R.J.; et al. An ingestible bacterial-electronic system to monitor gastrointestinal health. *Science* **2018**, *360*, 915–918. [CrossRef] [PubMed]
121. De la Paz, E.; Maganti, H.; Trifonov, A.; Jeerapan, I.; Mahato, K.; Yin, L.; Sonsa-ard, T.; Ma, N.; Jung, W.; Burns, R.; et al. A self-powered ingestible wireless biosensing system for real-time in situ monitoring of gastrointestinal tract metabolites. *Nat. Commun.* **2022**, *13*, 7405. [CrossRef] [PubMed]
122. Lee, Y.Y.; Erdogan, A.; Rao, S.S.C. How to assess regional and whole gut transit time with wireless motility capsule. *J. Neurogastroenterol. Motil.* **2014**, *20*, 265–270. [CrossRef] [PubMed]
123. Haffner, J.J.; Katemauswa, M.; Kagone, T.S.; Hossain, E.; Jacobson, D.; Flores, K.; Parab, A.R.; Obregon-Tito, A.J.; Tito, R.Y.; Reyes, L.M.; et al. Untargeted fecal metabolomic analyses across an industrialization gradient reveal shared metabolites and impact of industrialization on fecal microbiome-metabolome interactions. *mSystems* **2022**, *7*, e00710-22. [CrossRef] [PubMed]
124. Schaan, A.P.; Sarquis, D.; Cavalcante, G.C.; Magalhães, L.; Sacuena, E.R.P.; Costa, J.; Fonseca, D.; Mello, V.J.; Guerreiro, J.F.; Ribeiro-dos-Santos, A. The structure of Brazilian Amazonian gut microbiomes in the process of urbanisation. *npj Biofilms Microbiomes* **2021**, *7*, 65. [CrossRef] [PubMed]
125. Alberts, S.C.; Altmann, J. The Amboseli Baboon research project: 40 years of continuity and change. In *Long-Term Field Studies of Primates*; Kappeler, P., Watts, D.P., Eds.; Springer: Berlin/Heidelberg, Germany, 2012; pp. 261–287.
126. Adhyaru, B.B.; Jacobson, T.A. Safety and efficacy of statin therapy. *Nat. Rev. Cardiol.* **2018**, *15*, 757–769. [CrossRef] [PubMed]
127. Chen, Z.; Wu, P.; Wang, J.; Chen, P.; Fang, Z.; Luo, F. The association of statin therapy and cancer: A meta-analysis. *Lipids Health Dis.* **2023**, *22*, 192. [CrossRef] [PubMed]
128. Neogi, T.; Colloca, L. Placebo effects in osteoarthritis: Implications for treatment and drug development. *Nat. Rev. Rheumatol.* **2023**, *19*, 613–626. [CrossRef] [PubMed]

9. Benegens and the Closing of Pandora's Box

9.1. Holobiont Repair

The epidemiology of non-communicable disease suggests the need for the transmission of an intact, fully functioning microbiome from mother to offspring. The arguments set out in this book imply that the primary losses from the microbiome of people suffering from Burkitt's "Diseases of Modern Western Civilization" [1] are microbial sentinel cells, agents of the intergenerational immune system, whose purpose is to transfer antigenic information from the maternal microbiome to the neonate at the moment of birth. Accordingly, the health of future generations requires that these agents be re-introduced into the population, a process that can only be performed at the time of birth, just as the eventual adult immune system is first being assembled.

9.2. Benegens

In principle, such thus-far hypothetical microbial sentinel cells could be termed "birth probiotics", but their theoretical ability to provide benefits to future generations, and their potentially unique mode of action, suggests a new name: "benegens". Analysis further suggests that regular exposure to microbe-sized, sparingly soluble heavy-metal particles serves to deplete such cells, while oral antibiotics may have the effect of hindering the transfer of maternal microbiota to the child. Likewise, in the absence of the swabbing procedure known variously as microbirthing, vaginal seeding or vaginal inoculation, caesarean section may well prevent such transfer completely [2]. Accordingly, three steps will be needed to protect populations from further disease:

1. Identify and re-introduce microbial sentinel cells and other components of the fully functioning microbiome.
2. Limit opportunities for pollution by toxic heavy-metal ions while using bioremediation to reduce the toxic load currently in the environment.
3. Prevent further deterioration by modifying the policies surrounding C-section delivery and oral antibiotic use while ensuring the safety of the individual being treated.

9.3. Nature of Microbial Sentinel Cells

The analysis presented in this work strongly suggests that such cells have co-evolved with our own immune cells and will, therefore, be hard to distinguish. Indeed, it could be argued that benegen-like activity could follow if our own sentinel cells were sloughed off the wall of the intestine and transferred directly to the neonate. A counterargument to this suggestion is the expectation that these triple plagues would diminish if the toxic load were decreased. Sadly, nearly 30 years after the cessation of the use of leaded petrol [3], there is no sign that the population as a whole is regaining its health. Nevertheless, there remains a possibility that benegens and our own immune cells are one and the same.

9.4. Detection of Microbial Sentinel Cells

Perhaps the only way to definitively detect benegens is to catch them in action, actually transferring antigens from mother to child. In turn, perhaps the best way to do this is to use a harmless antigen-containing substance attached to a readily detected, harmless label, possibly including very low levels of radioactivity. The idea would be to swab the skin of the neonate immediately after birth. Of course, this dual inheritance hypothesis applies to all vertebrate animals, although our companion animals are exposed to the same toxins as their human handlers, and suffer from similar diseases [4]. Similarly, farmed animals are grown for their meat, the finer details of their long-term health being largely inconsequential. Likewise, laboratory animals are currently raised for their genes, while the loss of specific bacteria between different suppliers has already been documented [5]. The ideal animal subject may well be semi-wild animals, such as the baboons of the Amboseli National Park, Kenya, for example [6].

There is an increasing amount of work being carried out that involves people living in traditional societies and, with their agreement, this would be the best way to perform such studies. The primary drawback is the speed with which so-called urbanisation is taking place [7], probably with concomitant loss of microbial species [8]. Needless to say, any help provided by such peoples should be compensated for, possibly by the mechanisms developed for cancer treatment-related ethnobotanical knowledge [9]. Of course, it would be necessary to ensure that any such compensation does not itself destroy their microbiome and introduce non-communicable disease into their populations. For the ultimate benefit of humanity, Pandora's box must remain closed.

9.5. Funding

This work covers essentially all the elements of non-communicable disease, having the potential to draw all the threads together to solve these increasingly dangerous conditions. However, it is clear that major changes will be needed across society. I would suggest requesting a portion of every healthcare-related budget, including veterinary, in order to establish a committee dedicated to making recommendations in light of this dual inheritance hypothesis.

9.6. Changing the World

Interestingly, the whole world is now facing the same kind of challenges as did the Western world in the mid-1800s, except that the threat is now from non-communicable conditions, rather than infectious disease spreading through the overcrowded, unsanitary towns of the day. The two types of disease are by no means equivalent, however, as many survivors of infectious disease will bounce back to health, whereas those afflicted by these chronic conditions may even pass the problem on to subsequent generations. In those days, change followed the development of a new hypothesis, as the miasma theory of disease was eventually replaced by germ theory. One early attempt to change opinions was made by a Hungarian physician called Ignaz Semmelweis (1818–1865), who discovered a way to prevent the transmission of puerperal fever by the washing of hands with chlorine bleach. Although he had published sufficient evidence, his attempts to convince his colleagues failed completely, leading to be him becoming increasingly distraught until he met his death in a mental hospital [10]. It seems that, no matter how strong the evidence, real change requires peer-group consensus, without which nothing will happen. Nevertheless, Semmelweis' findings were widely reported, and he contributed to the changing zeitgeist as the 19th century wore on. At about the same time as Semmelweis was attempting to cajole his colleagues, Louis Pasteur (1822–1895) was completing his fermentation studies and realising their significance for infectious disease. Perhaps assisted by Semmelweis' efforts, Pasteur's ideas were picked up, and put into practice, by his younger colleagues, thus allowing consensus to develop over the longer term [11]. Even so, not everybody accepted that germs were all bad and, as we have seen, Élie Metchnikoff (1845–1916) was developing his ideas around the seemingly "good" Lactobacilli [12]. The ideas developed in this book derive from Lynn Margulis (1938–2011), suggesting that evolved, symbiotic microbes are neither exclusively "good", nor entirely "bad" [13]. Similarly, while Margulis championed the idea that microeukaryotes evolved by an exchange of genes between two prokaryote precursors, rendering neither capable of long-term survival [14], the same process may well have taken place prior to the evolution of the vertebrates. Of course, the science of microbiology developed in the context of pathogenicity, using what has become known as the Petri dish as the standard apparatus [15]. Although outside the scope of this work, in honour of Lynn Margulis it could be that the science of non-pathogenic/gut wall/body interactions should be described as *holobiology* [16].

References

1. Burkitt, D.P. Some diseases characteristic of modern western civilization. *Br. Med. J.* **1973**, *1*, 274–278. [CrossRef] [PubMed]
2. Dunn, A.B.; Jordan, S.; Baker, B.J.; Carlson, N.S. The maternal infant microbiome: Considerations for labor and birth. *MCN Am. J. Matern. Child Nurs.* **2017**, *42*, 318–325. [CrossRef] [PubMed]
3. Needleman, H. The removal of lead from gasoline: Historical and personal reflections. *Environ. Res.* **2000**, *84*, 20–35. [CrossRef] [PubMed]
4. Marsella, R. Atopic dermatitis in domestic animals: What our current understanding is and how this applies to clinical practise. *Vet. Sci.* **2021**, *8*, 124. [CrossRef] [PubMed]
5. Burberry, A.; Wells, M.F.; Limone, F.; Couto, A.; Smith, K.S.; Keaney, J.; Gillet, G.; van Gastel, N.; Wang, J.-Y.; Pietilainen, O.; et al. C9orf72 suppresses systemic and neural inflammation induced by gut bacteria. *Nature* **2020**, *582*, 89–94. [CrossRef] [PubMed]
6. Alberts, S.C.; Altmann, J. The Amboseli Baboon research project: 40 years of continuity and change. In *Long-Term Field Studies of Primates*; Kappeler, P., Watts, D.P., Eds.; Springer: Berlin/Heidelberg, Germany, 2012; pp. 261–287.

7. Schaan, A.P.; Sarquis, D.; Cavalcante, G.C.; Magalhães, L.; Sacuena, E.R.P.; Costa, J.; Fonseca, D.; Mello, V.J.; Guerreiro, J.F.; Ribeiro-dos-Santos, A. The structure of Brazilian Amazonian gut microbiomes in the process of urbanisation. *npj Biofilms Microbiomes* **2021**, *7*, 65. [CrossRef] [PubMed]
8. Reese, A.T.; Chadaideh, K.S.; Diggins, C.E.; Schell, L.D.; Beckel, M.; Callahan, P.; Ryan, R.; Thompson, M.E.; Carmody, R.N. Effects of domestication on the gut microbiota parallel those of human industrialization. *eLife* **2021**, *10*, e60197. [CrossRef] [PubMed]
9. Ryan, C.R. Towards an ethics of reciprocity: Ethnobotanical knowledge and medicinal plants as cancer therapies. *Humanities* **2014**, *3*, 624–644. [CrossRef]
10. Kadar, N. Rediscovering Ignaz Pilipp Semmelweis (1818–1865). *Am. J. Obstet. Gynecol.* **2019**, *220*, 26–39. [CrossRef] [PubMed]
11. Cavaillon, J.-M.; Legout, S. Louis Pasteur: Between myth and reality. *Biomolecules* **2022**, *12*, 596. [CrossRef] [PubMed]
12. Stambler, I. Elie Metchnikov. The founder of longevity science and a founder of modern medicine: In honour of the 170th anniversary. *Adv. Gerontol.* **2015**, *5*, 201–208. [CrossRef]
13. Guerrero, R.; Margulis, L.; Berlanga, M. Symbiogenesis: The holobiont as a unit of evolution. *Int. Microbiol.* **2013**, *16*, 133–143. [PubMed]
14. Laforest-Lapointe, I.; Arrieta, M.-C. Microbial eukaryotes: A missing link in gut microbiome studies. *mSystems* **2018**, *3*, e00201-17. [CrossRef] [PubMed]
15. Shama, G. The "Petri" dish: A case of simultaneous invention in bacteriology. *Endeavour* **2019**, *43*, 11–16. [CrossRef] [PubMed]
16. Margulis, L. Symbiogenesis and symbionticism. In *Symbiosis as a Source of Evolutionary Innovation: Speciation and Morphogenesis*; Margulis, L., Fester, R., Eds.; MIT Press: Cambridge, MA, USA, 1991; pp. 49–92.

MDPI AG
Grosspeteranlage 5
4052 Basel
Switzerland
Tel.: +41 61 683 77 34

MDPI Books Editorial Office
E-mail: books@mdpi.com
www.mdpi.com/books

www.ingramcontent.com/pod-product-compliance
Lightning Source LLC
Chambersburg PA
CBHW040852230926
51850CB00012BA/2935
* 9 7 8 3 7 2 5 8 1 5 2 4 1 *